Quantifying Climate Risk and Building Resilience in the UK

Suraje Dessai · Kate Lonsdale · Jason Lowe ·
Rachel Harcourt
Editors

Quantifying Climate Risk and Building Resilience in the UK

Editors
Suraje Dessai
University of Leeds
Leeds, UK

Kate Lonsdale
Climate Sense
Sheffield, UK

Jason Lowe
Met Office
Exeter, UK

Rachel Harcourt
University of Leeds
Leeds, UK

ISBN 978-3-031-39728-8   ISBN 978-3-031-39729-5 (eBook)
https://doi.org/10.1007/978-3-031-39729-5

© The Editor(s) (if applicable) and The Author(s) 2024. This book is an open access publication.

Open Access This book is licensed under the terms of the Creative Commons Attribution 4.0 International License (http://creativecommons.org/licenses/by/4.0/), which permits use, sharing, adaptation, distribution and reproduction in any medium or format, as long as you give appropriate credit to the original author(s) and the source, provide a link to the Creative Commons license and indicate if changes were made.
The images or other third party material in this book are included in the book's Creative Commons license, unless indicated otherwise in a credit line to the material. If material is not included in the book's Creative Commons license and your intended use is not permitted by statutory regulation or exceeds the permitted use, you will need to obtain permission directly from the copyright holder.
The use of general descriptive names, registered names, trademarks, service marks, etc. in this publication does not imply, even in the absence of a specific statement, that such names are exempt from the relevant protective laws and regulations and therefore free for general use.
The publisher, the authors, and the editors are safe to assume that the advice and information in this book are believed to be true and accurate at the date of publication. Neither the publisher nor the authors or the editors give a warranty, expressed or implied, with respect to the material contained herein or for any errors or omissions that may have been made. The publisher remains neutral with regard to jurisdictional claims in published maps and institutional affiliations.

Cover illustration: © Melisa Hasan

This Palgrave Macmillan imprint is published by the registered company Springer Nature Switzerland AG
The registered company address is: Gewerbestrasse 11, 6330 Cham, Switzerland

Foreword

Emission of greenhouse gases by human activity has now warmed the average temperature of our planet by more than 1°C, causing significant change to the climate system. It is critical that we limit further emissions and meet the international goals of the Paris Agreement but even as we do so, warming and climate change will continue. Human societies have developed with rather constant climate conditions, so today's changes are a shock to our systems. To thrive in the future we must be ready. We must be resilient to climate change.

In the UK we have seen increased instances of flooding, water shortages, storm damage, coastal denudation, and summer temperatures that have topped 40°C for the first time in recorded history. Modelled climate projections show us, with confidence, that these changes will become more severe even with the most ambitious emissions reductions. How we prepare for these changes will determine the scale of harm caused to our infrastructure, our health, the systems we rely on such as food and transport systems, and the harm we inflict on nature.

The UK has been a global leader in climate adaptation policy. The Climate Change Act 2008 legally requires the UK government to conduct a five-yearly Climate Change Risk Assessment (CCRA) for the UK, and requires my Government Department—Defra—to produce a National Adaptation Programme for England. Devolved legislation provides a comparable framework of adaptation planning and evaluation for Scotland, Wales, and Northern Ireland. UK research has also been world leading in identifying actions needed to manage climate risks and to reduce damage without exacerbating existing inequalities. There is much more to do, however, and national preparedness against many of the climate risks identified in the last CCRA remains inadequate. Governments, business, and wider society need to plan and take action to improve our resilience to climate change.

The *UK Climate Resilience Programme* (UKCR) is a research programme addressing this need for adaptation action. It was funded by UK Research and Innovation with strong support from Defra throughout its work. UKCR enhanced the UK's resilience to climate variability and change through interdisciplinary research and innovation on climate risk, adaptation and climate services, working with stakeholders and end users to ensure the research is useful and usable.

UKCR made significant contributions to the UK's third CCRA, and its tools, datasets, and learning will be important for the UK's fourth CCRA and beyond. The programme has deepened our understanding of the climate services sector, with projects on climate service standards and valuation, and a roadmap for development and implementation of UK climate services. Engagement with cities across the UK has raised awareness in a wide range of audiences about the need to respond to the climate challenge.

UKCR has not all been about natural science, but has included valuable contributions from many disciplines. The programme funded arts and community-based projects, and pioneered an embedded researcher scheme in which researchers collaborate with host organisations to help them understand and address their needs. The programme has been hugely valuable in creating a more coherent community of climate resilience researchers and practitioners across the UK.

I congratulate the UKCR programme for their achievements, captured in this book. Their learning and communication will be of long-lasting value to the climate resilience community. I hope that the climate resilience community that came together to achieve this work will go from strength to strength as we work together to build our resilience in the face of climate change.

Professor Gideon Henderson
Chief Scientific Advisor
Department for Environment
Food and Rural Affairs
London, UK

Acknowledgements

This work was supported by the UK Research and Innovation and Met Office as part of the UK Climate Resilience Programme (UKCR) [grant number NE/S017321/1]; the Economic and Social Research Council (ESRC) as part of the work of the ESRC Centre for Climate Change Economics and Policy [ES/R009708/1]. We also thank members of the UKCR Programme Board and the Steering Committee, chaired by Professor Jim Hall.

Many thanks to our lead and contributing authors who provided their time and commitment to creating this synthesis of the UKCR programme. We would also like to thank the peer reviewers of the main chapters for their engaged and insightful suggestions.

Special thanks to Catherine Homoky, Kate Lock, Pete Walton, Zorica Jones, and Nick Hopkins-Bond for their support in developing this book. Thanks also go to the whole UKCR Champion team based at the University of Leeds, as well as UKRI-based and Met Office-based UKCR colleagues.

Finally, we thank everyone who has supported and contributed to the programme through delivering research, participating in events, and ensuring the UK is building resilience to a changing climate.

Contents

1 Introducing the UK Climate Resilience Programme　　1
   Suraje Dessai, Kate Lonsdale, Jason Lowe,
   and Rachel Harcourt

2 Climate Resilience: Interpretations of the Term
   and Implications for Practice　　15
   Kate Lonsdale, Nigel Arnell, Tim Coles, Kate Lock,
   Emer O'Connell, Paul O'Hare, and Emma Tompkins

   Part I　Undertaking Resilience Research

3 Towards a Step Change in Co-Production for Climate
   Resilience　　27
   Nicola Golding, Jenna Ashton, Kate Brown,
   Steven Chan, Tim Coles, Hayley Fowler,
   Elizabeth Fuller, Paula Harrison,
   Alice Harvey-Fishenden, Neil Macdonald,
   and Christine Sefton

4 Learning from Organisational Embedding for Climate
   Resilience　　43
   Kate Lonsdale, Tim Coles, Paul O'Hare, Caitlin Douglas,
   Stephen Scott-Bottoms, Alan Kennedy-Asser,
   Charles Rougé, and Corinna Wagner

Part II Managing Climate Risks

5 **Putting Climate Resilience in Its Place: Developing Spatially Literate Climate Adaptation Initiatives**   63
Freya Garry, Paul O'Hare, Claire Scannell, Jenna Ashton, Michael Davies, Katy A. Freeborough, Alan Kennedy-Asser, Neil Macdonald, Stephen Scott-Bottoms, and Liz Sharp

6 **Learning from Arts and Humanities Approaches to Building Climate Resilience in the UK**   75
Edward Brookes, Briony McDonagh, Corinna Wagner, Jenna Ashton, Alice Harvey-Fishenden, Alan Kennedy-Asser, Neil Macdonald, and Kate Smith

Part III Tools for Resilience Building

7 **What Have We Learned from the Climate Service Projects Delivered Through the UK Climate Resilience Programme?**   93
Caitlin Douglas and Mark Harrison

8 **What Insights Can the Programme Share on Developing Decision Support Tools?**   111
Rachel Perks, Craig Robson, Nigel Arnell, James Cooper, Laura Dawkins, Elizabeth Fuller, Alan Kennedy-Asser, Robert Nicholls, and Victoria Ramsey

Part IV Understanding and Characterising Risk

9 **Improved Understanding and Characterisation of Climate Hazards in the UK**   131
Jennifer Catto, Simon Brown, Clair Barnes, Steven Chan, Daniel Cotterill, Murray Dale, Laura Dawkins, Hayley Fowler, Freya Garry, Will Keat, Elizabeth Kendon, Jason Lowe, Colin Manning, David Pritchard, Peter Robins, David Sexton, Rob Shooter, and David Stephenson

10	Future Changes in Indicators of Climate Hazard and Resource in the UK	145
	Nigel Arnell, Stephen Dorling, Hayley Fowler, Helen Hanlon, Katie Jenkins, and Alan Kennedy-Asser	
11	What Has Been Learned About Converting Climate Hazard Data to Climate Risk Information?	163
	Dan Bernie, Freya Garry, Katie Jenkins, Nigel Arnell, Laura Dawkins, Alistair Ford, Alan Kennedy-Asser, Paul O'Hare, Rachel Perks, Victoria Ramsey, and Paul Sayers	
12	Note on Delivering Impact	177
	Rachel Harcourt and Nick Hopkins-Bond	
13	Afterword	185
	Suraje Dessai, Kate Lonsdale, Jason Lowe, and Rachel Harcourt	
Project References		201
Index		215

List of Contributors

Nigel Arnell University of Reading, Reading, UK

Jenna Ashton University of Manchester, Manchester, UK

Clair Barnes University College London, London, UK

Dan Bernie Met Office, Exeter, UK; University of Bristol, Bristol, UK

Edward Brookes University of Hull, Hull, UK

Kate Brown Met Office, Exeter, UK

Simon Brown Met Office, Exeter, UK

Jennifer Catto University of Exeter, Exeter, UK

Steven Chan Newcastle University, Newcastle upon Tyne, UK

Tim Coles University of Exeter, Exeter, UK

James Cooper University of Liverpool, Liverpool, UK

Daniel Cotterill Met Office, Exeter, UK

Murray Dale JBA Consulting, Exeter, UK

Michael Davies University College London, London, UK

Laura Dawkins Met Office, Exeter, UK

xvi LIST OF CONTRIBUTORS

Suraje Dessai University of Leeds, Leeds, UK

Stephen Dorling University of East Anglia, Norwich, UK

Caitlin Douglas King's College London, London, UK

Alistair Ford Newcastle University, Newcastle upon Tyne, UK

Hayley Fowler Newcastle University, Newcastle upon Tyne, UK

Katy A. Freeborough British Geological Survey, Nottingham, UK

Elizabeth Fuller Met Office, Exeter, UK

Freya Garry Met Office, Exeter, UK

Nicola Golding Met Office, Exeter, UK

Helen Hanlon Met Office, Exeter, UK

Rachel Harcourt University of Leeds, Leeds, UK

Mark Harrison Met Office, Exeter, UK

Paula Harrison Centre for Ecology & Hydrology, Lancaster, UK

Alice Harvey-Fishenden University of Liverpool, Liverpool, UK

Nick Hopkins-Bond Met Office, Exeter, UK

Katie Jenkins University of East Anglia, Norwich, UK

Will Keat Met Office, Exeter, UK

Elizabeth Kendon Met Office, Exeter, UK; University of Bristol, Bristol, UK

Alan Kennedy-Asser University of Bristol, Bristol, UK

Kate Lock University of Leeds, Leeds, UK

Kate Lonsdale Climate Sense, Sheffield, UK

Jason Lowe Met Office, Exeter, UK

Neil Macdonald University of Liverpool, Liverpool, UK

Colin Manning Newcastle University, Newcastle upon Tyne, UK

Briony McDonagh University of Hull, Hull, UK

Robert Nicholls University of East Anglia, Norwich, UK

Emer O'Connell Greater London Authority, London, UK

Paul O'Hare Manchester Metropolitan University, Manchester, UK

Rachel Perks Met Office, Exeter, UK

David Pritchard Newcastle University, Newcastle upon Tyne, UK

Victoria Ramsey Met Office, Exeter, UK

Peter Robins Bangor University, Bangor, Wales, UK

Craig Robson Newcastle University, Newcastle upon Tyne, UK

Charles Rougé University of Sheffield, Sheffield, UK

Paul Sayers University of East Anglia, Norwich, UK; Sayers and Partners, Watlington, UK

Claire Scannell Met Éireann, Dublin, Ireland

Stephen Scott-Bottoms University of Manchester, Manchester, UK

Christine Sefton University of Sheffield, Sheffield, UK

David Sexton Met Office, Exeter, UK

Liz Sharp University of Sheffield, Sheffield, UK

Rob Shooter Met Office, Exeter, UK

Kate Smith University of Hull, Hull, UK

David Stephenson University of Exeter, Exeter, UK

Emma Tompkins University of Southampton, Southampton, UK

Corinna Wagner University of Exeter, Exeter, UK

List of Figures

Chapter 1

Fig. 1 Envisioned programme legacies identified in the UKCR science plan 10

Chapter 3

Fig. 1 Summary of the key themes emerging from UKCR projects relating to a required step change in co-production research for climate resilience 37

Chapter 6

Fig. 1 Appledore Time and Tide Bell. Artist: Marcus Vergette (*Photograph* Corinna Wagner, 2021) 80
Fig. 2 Creating objects for the 'Creative Climate Resilience' project (*Photograph* Jenna Ashton, 2021) 82
Fig. 3 Audience members enjoying the Sinuous City installation, part of the FloodLights event in Hull (*Photograph* Briony McDonagh, 2021) 85

Chapter 7

Fig. 1   The enabling environment in which a prototype is being developed affects its development and delivery. Conditions within the provider organisation, the user organisation and the wider context in which they operate all need to incentivise climate service delivery   101

Chapter 8

Fig. 1   From the UKCR project 'Once Upon a Time', an example of an interactive DST, which allows users to explore the changes in temperature rise across Northern Ireland over time (*Source* https://akaresearch.shinyapps.io/ruralheat/)   118

Fig. 2   Example pages from the Heat Pack for Bristol, one of the products of the 'Heat Service' (Meeting Urban User Needs) project. These PDFs are designed to inform stakeholders on possible effects of climate change to help inform decision making (*Source* https://www.metoffice.gov.uk/binaries/content/assets/metofficegovuk/pdf/research/spf/ukcr_heat_pack_bristol.pdf)   120

Chapter 9

Fig. 1   Return period diagram (From Leach et al. [11]) showing the return period of UK DJF maximum surface temperature anomaly from the UKCP ensemble from 2061–2080. The black curve shows the median of the generalised extreme value (GEV) model fit and the dotted lines show the 0.1–99.9% confidence interval on the GEV fit. The thin orange lines on the left show the UK DJF maximum surface temperature anomalies from the ExSamples ensemble. This figure demonstrates that using the relatively low-resolution model in this study, forced with the SSTs from an extreme hot winter, produces even larger anomalies than the UKCP ensemble. Reproduced according to the CC-BY licence   135

Chapter 11

Fig. 1 Risk as a function of hazard, exposure, vulnerability and response. This example illustrates some of the complex interactions that generated risk to infrastructure during the 2018 European heatwave [4]   165

Chapter 12

Fig. 1 Best practice principles from the UKCR programme for delivering impact   179

Chapter 13

Fig. 1 UKCR's knowledge brokering, translation and application roles building on [14–16]   192

LIST OF TABLES

Chapter 4

Table 1   The details of the funded projects across the two cohorts   47

Chapter 7

Table 1   Climate services projects commissioned by the UK Climate Resilience Programme. We define 'applied research' as capability-led development with subsequent user engagement, 'prototype development' as exploratory work to develop new products for new markets, 'product development' as the creation of new products in response to clear user needs, and 'delivery' as being additional activities to support the development and delivery of climate services. We define sector maturity as: low (no regulation), medium (regulation but minimal engagement), or high (active engagement with regulation)   96

Chapter 8

Table 1   A summary of the projects interviewed for the survey of UKCR decision support tools, including a description of each tool, the spatial scale it operates on and the stakeholders directly involved in its development   114

Chapter 9

Table 1　The table provides UKCR outputs relating to hazards, as described in this chapter: tools and code (blue); websites for data exploration (yellow); and datasets (green)　134

Chapter 10

Table 1　Changes in indicators of climate hazard and resource by the 2050s, based on the central estimate from the HadGEM UKCP18 strand (global, regional or local), with very high RCP8.5 emissions　149

Chapter 11

Table 1　A selection of the new datasets for hazard, vulnerability, exposure and risk developed through the UKCR programme　170

CHAPTER 1

Introducing the UK Climate Resilience Programme

Suraje Dessai, Kate Lonsdale, Jason Lowe and Rachel Harcourt

Abstract

- Research and policy relating to climate change risks and adaptation have been developing in the UK since the 1990s.

S. Dessai (✉) · K. Lonsdale · J. Lowe · R. Harcourt
University of Leeds, Leeds, UK
e-mail: s.dessai@leeds.ac.uk

K. Lonsdale
e-mail: kate.lonsdale@climatesense.global

J. Lowe
e-mail: jason.lowe@metoffice.gov.uk

R. Harcourt
e-mail: r.s.harcourt@leeds.ac.uk

J. Lowe
Met Office, Exeter, UK

© The Author(s) 2024
S. Dessai et al. (eds.), *Quantifying Climate Risk and Building Resilience in the UK*,
https://doi.org/10.1007/978-3-031-39729-5_1

- The 2008 Climate Change Act established much of the framework for UK climate risk assessment and adaptation management.
- The UK Climate Resilience Programme was funded from 2018–2023 to address research gaps in characterising and quantifying climate risks, managing climate related risks through adaptation, and co-producing climate services.
- From the outset, the programme prioritised co-production, innovation, trans-disciplinary research, and working with stakeholders to ensure outputs were useful and usable.

Keywords Climate risks · Adaptation · Resilience · Policy

1 Introduction

Global and UK climate is changing at an unprecedented rate. New weather and climate records are being set, and there is growing evidence that human activity is influencing the probability of dangerous climate extremes [5]. Further climate change is now inevitable, but the amount and pace of change will be shaped by the effectiveness of international climate mitigation policies. Recent past and projected future change mean that adaptation is critical in reducing climate risk and vulnerability in human and natural systems. The concepts of adaptation, vulnerability, resilience and risk provide overlapping, alternative entry points for the climate change challenge (see box of key terms).

> **Key terms [7]**
>
> **Mitigation** (of climate change): A human intervention to reduce emissions or enhance the sinks of greenhouse gases.
> **Adaptation**: In human systems, the process of adjustment to actual or expected climate and its effects, in order to moderate harm or exploit beneficial opportunities. In natural systems, the process of adjustment to actual climate and its effects; human intervention may facilitate adjustment to expected climate and its effects.

Risk: The potential for adverse consequences for human or ecological systems, recognising the diversity of values and objectives associated with such systems. In the context of climate change, risks can arise from potential impacts of climate change as well as human responses to climate change. In the context of climate change impacts, risks result from dynamic interactions between climate-related hazards with the exposure and vulnerability of the affected human or ecological system to the hazards.

Hazard: The potential occurrence of a natural or human-induced physical event or trend that may cause loss of life, injury or other health impacts, as well as damage and loss to property, infrastructure, livelihoods, service provision, ecosystems and environmental resources.

Exposure: The presence of people; livelihoods; species or ecosystems; environmental functions, services, and resources; infrastructure; or economic, social, or cultural assets in places and settings that could be adversely affected.

Vulnerability: The propensity or predisposition to be adversely affected. Vulnerability encompasses a variety of concepts and elements, including sensitivity or susceptibility to harm and lack of capacity to cope and adapt.

Impacts: The consequences of realised risks on natural and human systems, where risks result from the interactions of climate-related hazards (including extreme weather/climate events), exposure, and vulnerability.

Resilience: The capacity of interconnected social, economic and ecological systems to cope with a hazardous event, trend or disturbance, responding or reorganising in ways that maintain their essential function, identity and structure. Resilience is a positive attribute when it maintains capacity for adaptation, learning and/or transformation.

Projection: A potential future evolution of a quantity or set of quantities, often computed with the aid of a model. Unlike predictions, projections are conditional on assumptions concerning, for example, future socio-economic and technological developments that may or may not be realised.

> **Scenario**: A plausible description of how the future may develop based on a coherent and internally consistent set of assumptions about key driving forces and relationships.

In the UK, the decade 2012–2021 was 1°C warmer than the 1961–1990 average, compared with 0.8°C for global surface temperature [10]. In July 2022, the UK experienced unprecedented high temperatures above 40°C and there is growing evidence that the intensity of heavy rainfall events has increased in recent years [4]. Changes in UK climatic impact-drivers—the climate conditions that affect the things we care about in nature and society [15]—have led to a multitude of climate impacts, such as infrastructure damage and heat-related deaths. There is some evidence of adaptation action in the UK [8], which could offset some of the increased risk due to climate change, but the UK's Climate Change Committee states that the gap between the level of risk and the level of adaptation has widened recently for many sectors [3].

The UK research base in climate risk and resilience is world leading but fragmented. Climate scientists at the Met Office and in UK universities have been at the forefront of climate science and climate services, exemplified by the UK Climate Projections 2018 (UKCP18) and the development of high-resolution global climate modelling. Engineering and other sciences have translated climate hazard knowledge into impact and risk metrics (for example, in infrastructure and urban adaptation), creating a national infrastructure system-of-systems model. Social science research has focused on areas such as barriers to adaptation, economic costs and benefits, risk perception, behaviour and communication, and the science-policy interface. The arts and humanities have achieved contributions in the philosophy of climate science, the history, heritage, ethics and culture of climate change, climate adaptation and resilience, and artistic interventions focusing on living with change and loss. The UK was also world leading in setting up a 'boundary organisation' to act as a bridge between scientific research, policymaking and adaptation practice through the UK Climate Impacts Programme (UKCIP) in 1997. The next section briefly reviews key developments in UK climate adaptation research and policy.

2 A Brief History of UK Climate Research and Policy on Adaptation

In the 1990s, the UK government founded the Met Office Hadley Centre (1990) [12], published the first national assessment of the possible impacts of climate change (1991) [4] and established the pioneering UK Climate Impacts Programme (UKCIP) to bring together scientific research, policymaking and adaptation practice (1997) [13]. While UKCIP's overarching aim was to help the UK adapt to the unavoidable impacts of climate change, its remit shifted over time—from engaging organisations on initial impact assessment, to occupying the boundary space between climate projections and research, supporting policy development (including the legislative requirements of the Climate Change Act 2008) and helping organisations develop and implement adaptation strategies and actions. As well as developing a portfolio of tools, UKCIP published the first national climate change scenarios in 1998 and was influential in framing climate adaptation as a problem of risk management [19].

During the 2000s, more national climate change scenarios were published by the Met Office (2001 and 2009) and the UK developed a regional, multi-sectoral and integrated assessment of the impacts of climate and socioeconomic change in the UK [9]. The Climate Change Act 2008 created a framework for adaptation to climate change, by establishing:

- the five-yearly, UK-wide Climate Change Risk Assessment (CCRA);
- that a National Adaptation Programme (NAP), must be put in place to address climate change risks and be reviewed following each CCRA;
- the 'Adaptation Reporting Power' (ARP), giving the government discretionary power to require relevant bodies to report on climate preparedness; and
- the Adaptation Sub-Committee (now the Adaptation Committee) of the independent climate change committee (CCC), to advise government and evaluate adaptation progress.

The UK government has since published three CCRAs (2012, 2017 and 2022), three NAPs for England (2013, 2018 and 2023) and there have been three rounds of ARP reporting. Responsibility for climate change adaptation is split between the four countries of the UK. The UK government is responsible for climate change adaptation in England and for reserved matters, with national governments in Northern Ireland, Wales and Scotland being responsible for adaptation in all devolved policy areas. The Adaptation Committee of the CCC has assessed progress of the UK and devolved governments in preparing for and adapting to the impacts of climate change. Between 2008 and 2010, local authorities reported against a process-based framework to help their preparations for a changing climate (known as National Indicator 188).

The early 2010s saw the development of regional climate change partnerships under the umbrella 'Climate UK' (for example, the London Climate Change Partnership and Climate Northern Ireland) and the transfer of many of UKCIP's functions to the Environment Agency's Climate Ready Programme in 2012. Financial austerity in the public sector led to the closure of most of the regional climate change partnerships and in March 2016, the Climate Ready Programme also ended, leaving England largely devoid of a boundary organisation focused on climate impacts and adaptation (with the exception of the Marine Climate Change Impacts Partnership).[1] However, since 2019 more than ten city and regional Climate Commissions have been established under the Place-based Climate Action Network (PCAN), working as independent advisory groups bringing together the public, private and civic sectors. During this period, research continued including the following programmes: the Programme of Research on Preparedness, Adaptation and Risk (PREPARE; 2012–2013); the Living With Environmental Change (LWEC) Programme (which included 'climate change impact' report cards and Engineering and Physical Sciences Research Council's (EPSRC) 'Adaptation and Resilience to a Changing Climate' (ARCC), amongst others); the Met Office Hadley Centre Climate Programme and the UK Climate Projections 2018 [11]; CCRA2 [1], and some decision support tools (for example, the Climate Just tool).

While undertaking the delivery of the CCRA2 evidence report, the then-named Adaptation Sub-Committee collated over 200 evidence gaps

---

[1] Boundary organisations continue to exist in Scotland (Adaptation Scotland) and Northern Ireland (Climate Northern Ireland).

which were discussed and supplemented at a research needs conference in late 2016. The committee concluded that to progress understanding of climate risks in future CCRAs, the following cross-cutting evidence gaps needed to be addressed:

- UK spatial modelling capability;
- Socioeconomic scenarios for the UK;
- Decision support frameworks;
- Monitoring;
- Behaviour change;
- Adaptation options.

In 2017, the chair of the Adaptation Committee wrote to Research Council Chief Executives stating that a concerted multidisciplinary effort was required to support future CCRAs. In mid-2018, with support from Department for Environment, Food and Rural Affairs (Defra)'s Chief Scientist, the Natural Environment Research Council (NERC) and the Met Office jointly submitted a proposal for a programme on Climate Resilience to UK Research and Innovation (UKRI)'s Wave 1 Strategic Priorities Fund (SPF). The SPF UK Climate Resilience (UKCR) programme was approved in the autumn of 2018 at a total cost of £18.65 million over the period 2018–2023, as a partnership between UKRI and the Met Office.

3 THE SCIENCE PLAN AND ITS IMPLEMENTATION

The Strategic Priorities Fund (SPF) offered an opportunity to improve climate risk assessment and enhance UK resilience by encouraging and funding high-quality multi- and interdisciplinary research and innovation using integrative approaches that cross traditional disciplinary boundaries. It provided space for pioneering research, laying the foundation for future capability, and aimed to link effectively with government departments' research priorities and opportunities. The UKCR programme is an example of SPF's ability to respond with agility to strategic priorities and opportunities, and ensure the UK remains at the cutting edge of research.

The vision for the UKCR programme was:

To enhance the UK's resilience to climate variability and change through frontier interdisciplinary research and innovation on climate risk, adaptation and climate services, working with stakeholders and end users to ensure the research is useful and usable.

The programme's three main objectives were:

- Characterising and quantifying climate-related risks;
- Managing climate-related risks through adaptation;
- Co-producing climate services.

The science plan recognised that single disciplinary approaches will not be able to 'solve' this complex challenge and that multi- and interdisciplinary research efforts that include the natural sciences, social sciences, engineering, the arts and humanities are needed. It also recognised that the engagement and involvement of a wider range of stakeholders, such as practitioners and policymakers, are essential in addressing this challenge.

A programme board was established to oversee UKCR investment, with representation from the following funding bodies: the Met Office, NERC, AHRC (Arts and Humanities Research Council), ESRC (Economic and Social Research Council) and EPSRC. An independent steering committee was also established, to provide strategic input on the shape and delivery of the initiative, policy, alignment with other initiatives and opportunities for programme coordination and coherence. Following a networking workshop in September 2018, NERC/UKRI issued an interdisciplinary call to supplement ongoing UKRI Research Council awards/activities in climate resilience,[2] and a further call for the role of UKCR champion. During this first round of funding, 19 projects were funded by NERC/UKRI, and Professor Suraje Dessai and Dr Kate Lonsdale, based at the University of Leeds, were appointed as champions to act as thought leaders, flag bearers and strategy owners for UKCR. The

---

[2] Funding projects up to £250k for up to 12 months with a total budget up to £3.5m.

champions worked closely with the Met Office to ensure integration of the programme and development of strategy.

The champions and the Met Office led the development of a joint science plan [16], in consultation with UKRI and Met Office communities and climate resilience stakeholders from public, private and third sectors, while also taking into consideration government departments' research priorities, the evolving UK climate resilience research landscape and the state-of-the-art in relevant disciplines. The science plan identified opportunities to significantly improve capabilities and address the challenges of quantifying risk and enhancing resilience. It was delivered through four main activities: frontier research, building research capability, developing and testing climate services, and coordination and networking activities. A central aspiration of the programme was to grow the community of interacting researchers, practitioners and policymakers in climate resilience. This underpinned all activities, but was particularly important in climate services, networking activities and the embedded researcher scheme. Guided by the science plan, UKRI issued seven open calls[3] and the Met Office commissioned 13 external projects[4] and 16 internal projects,[5] resulting in more than 60 projects funded across the whole programme. The science plan envisioned a series of legacies for the programme, as shown in Fig. 1:

Building on the science plan, the programme developed a high-level narrative to link the outlined 'legacy items' with UKCR's vision and objectives, articulating the activities needed to achieve the agreed legacy items, which would ultimately contribute to the programme goal of "enhancing the UK's resilience to climate variability and change". This narrative underpinned the monitoring, evaluation and learning plan for the programme. Programme activities were designed with consideration of the spectrum of knowledge brokering approaches, from 'linear

---

[3] UK Climate resilience first call; Champion; Enhancing climate change risk assessment capability; Governing Adaptation; AHRC Living with climate uncertainty; Present and future climate hazard; Embedded researchers.

[4] E.g. Development and provision of UK socioeconomic scenarios for climate vulnerability, impact, adaptation and services research and policy; Enabling the use and producing improved understanding of EuroCORDEX data over the UK; Climate services standards monitoring and valuing.

[5] Within four work packages: Improving climate hazard information; From climate hazard to risk; Climate services pilot; Operational climate services.

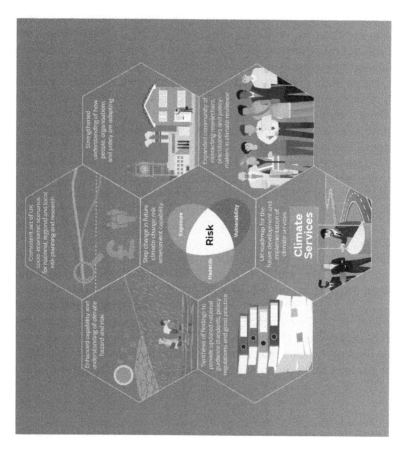

Fig. 1 Envisioned programme legacies identified in the UKCR science plan

dissemination of knowledge' to 'co-production'. Given the goal of the programme, traditional evaluation criteria, such as academic publications and citations, were expanded to include knowledge utilisation, knowledge exchange processes, and benefit of research to society. Thus, in addition to research excellence, the programme has assessed partnership and co-production, research relevance for target users, positioning of research outputs for use, and progress towards building a coherent climate resilience research community [11].

4 Book Roadmap

This book synthesises research conducted throughout the programme, through a series of chapters authored by UKCR researchers (usually a pairing from the Met Office and academia). At the time of writing in late 2022, several UKCR-funded projects were still ongoing, so some outputs have not been captured. Chapter 2 considers the key interpretations of climate resilience and its implications for practice. The rest of the chapters are organised in four parts.

In 'Part 1: Undertaking Resilience Research', we explore two means used in the programme to deliver context-specific, multi-stakeholder resilience research and achieve outputs suitable for practice and policy. Both co-production and embedding highlight the importance of more integrative approaches to climate resilience research, and this section provides guidance on how to deliver these approaches.

In 'Part 2: Managing Climate Risks', we focus on the place-based and context-specific nature of climate resilience research. Chapter 5 draws on projects that worked at specific geographical scales, to consider how connection to 'place' adds motivation and meaning to building climate resilience. Chapter 6 synthesises projects which directly interacted with *people* in particular places, using creative means to increase local engagement in discussions on climate risks. Both chapters emphasise the need for a 'context first' approach to climate resilience research and decision-making.

In 'Part 3: Tools for Resilience Building', we discuss climate services in Chapter 7 and decision support tools in Chapter 8. As well as summarising the contributions of the programme and highlighting research priorities going forward, both chapters consider what has been learned about the wider policy and practice context needed for the development and upscaling of climate services and decision-support tools.

Finally, in 'Part 4: Understanding and Characterising Risk', we summarise and signpost the programme's contributions to this area, including new methods, data sets and tools. Chapter 9 outlines how the programme used new tools to improve the projection of hazards and discusses how they can be used to inform decision-making. Chapter 10 summarises how projects used UKCP18 datasets to calculate how climate change is likely to affect climate-related hazards and resources in the UK. Chapter 11 considers how UKCR projects have contributed to developing hazard information into risk information, while also highlighting the need for improved exposure and vulnerability data and better understanding of compound, cascading and systemic risks.

In the Afterword, the editors summarise and reflect on the research undertaken by the UKCR programme and conclude with a series of key learnings and priorities for future research.

NB: In all chapters, the authors refer to UKCR-funded projects by their abbreviated titles. Please turn to the List of Projects section at the back of the book for a brief description of each project and the research team, plus website links. More information about the programme, its objectives, legacy items and funded research can be found at https://www.ukclimateresilience.org/.

References

1. Adger, W. N., Brown, I. and Surminski, S. 2018. Advances in risk assessment for climate change adaptation policy. *Philosophical Transactions of the Royal Society A* **376**(2121).
2. Brown, A., Gawith, M., Lonsdale, K. and Pringle, P. 2011. Managing adaptation: linking theory and practice. Oxford, UK, UK Climate Impacts Programme.
3. Climate Change Committee. 2021. *Independent Assessment of UK Climate Risk. Advice to Government for the UK's third Climate Change Risk Assessment (CCRA3)*. [Online]. Available at: Independent Assessment of UK Climate Risk - Climate Change Committee (theccc.org.uk).
4. UK Climate Change Impacts Review Group (CCIRG). 1991. The potential effects of climate change in the United Kingdom. London, Department of the Environment: 124.
5. Cotterill, D., Stott, P., Christidis, N. and Kendon, E. 2021. Increase in the frequency of extreme daily precipitation in the United Kingdom in autumn. *Weather and Climate Extremes*, **33**.
6. Intergovernmental Panel on Climate Change (IPCC). 2021. *Climate Change 2021: The Physical Science Basis. Contribution of Working Group I to the Sixth Assessment Report of the Intergovernmental Panel on Climate*

Change [Masson-Delmotte, V., P. Zhai, A. Pirani, S.L. Connors, C. Péan, S. Berger, N. Caud, Y. Chen, L. Goldfarb, M.I. Gomis, M. Huang, K. Leitzell, E. Lonnoy, J.B.R. Matthews, T.K. Maycock, T. Waterfield, O. Yelekçi, R. Yu, and B. Zhou (eds.)]. Cambridge University Press, Cambridge, United Kingdom and New York, NY, USA.
7. IPCC, V. Möller, R. v. Diemen, J. B. R. Matthews, C. Méndez, S. Semenov, J. S. Fuglestvedt and A. Reisinger 2022. Annex II: Glossary. In: *Climate Change 2022: Impacts, Adaptation, and Vulnerability. Contribution of Working Group II to the Sixth Assessment Report of the Intergovernmental Panel on Climate Change* [H.-O. Pörtner, D.C. Roberts, M. Tignor, E.S. Poloczanska, K. Mintenbeck, A. Alegría, M. Craig, S. Langsdorf, S. Löschke, V. Möller, A. Okem, B. Rama (eds.)]. Cambridge University Press, Cambridge, UK and New York, NY, USA, pp. 2897–2930.
8. Jenkins, K., Ford, A., Robson, C. and Nicholls, R.J. 2022. Identifying adaptation 'on the ground': Development of a UK adaptation Inventory. *Climate Risk Management*, **36**.
9. Holman, I. P., Nicholls, R.J., Berry, P.M., Harrison, P.A., Audsley, E., Shackley, S. and Rounsevell, M.D.A. 2005. A regional, multi-sectoral and integrated assessment of the impacts of climate and socio-economic change in the UK: Part II. Results. *Climatic Change* **71**(1), pp. 43–73.
10. Kendon, M., McCarthy, M., Jevrejeva, S., Matthews, A., Sparks, T., Garforth, J. and Kennedy, J. 2022. State of the UK Climate 2021. *International Journal of Climatology* **42**(S1), pp. 1–80.
11. Lowe, J.A., Bernie, D., Bett, P., Bricheno, L., Brown, S., Calvert, D., Clark, R., Eagle, K., Edwards, T., Fosser, G., Fung, F., Gohar, L., Good, P., Gregory, J., Harris, G., Howard, T., Kaye, N., Kendon, E., Krijnen, J., Maisey, P., McDonald, R., McInnes, R., McSweeney, C., Mitchell, J.F.B., Murphy, J., Palmer, M., Roberts, C., Rostron, J., Sexton, D., Thornton, H., Tinker, J., Tucker, S., Yamazaki, K., and Belcher, S. 2018. UKCP18 Science Overview Report. [Online]. Available at: UKCP18-Overview-report.pdf (metoffice.gov.uk).
12. Mahony, M. and Hulme, M. 2016. Modelling and the Nation: Institutionalising Climate Prediction in the UK, 1988–92. *Minerva* **54**(4), pp. 445–470.
13. Hedger, M., Connell, R. and Bramwell, P. 2006. Bridging the gap: empowering decision-making for adaptation through the UK Climate Impacts Programme. *Climate Policy* **6**, pp. 201–215.
14. Ofir Z, Schwandt, T., Duggan, C. and McLean, R. 2016. Research Quality Plus: A Holistic Approach to Evaluating Research, *International Development Research Centre*. [Online]. Available at: https://idl-bnc-idrc.dspacedirect.org/handle/10625/56428

15. Ranasinghe, R., Ruane, A.C., Vautard, R., Arnell, N., Coppola, E., Cruz, F.A., Dessai, S., Islam, A.S., RahimiM., Ruiz Carrascal, D., Sillmann, J., M. B. Sylla, Tebaldi, C., Wang, W. and Zaaboul, R. 2021. Climate Change Information for Regional Impact and for Risk Assessment. *Climate Change 2021: The Physical Science Basis. Contribution of Working Group I to the Sixth Assessment Report of the Intergovernmental Panel on Climate Change.* V. Masson-Delmotte, P. Zhai, A. Pirani et al. Cambridge, United Kingdom and New York, NY, USA, Cambridge University Press, pp. 1767–1926.
16. Ruane, A. C., Vautard, R., Ranasinghe, R., Sillmann, J., Coppola,E., Arnell, N., Cruz, F.A., Dessai, S., Iles, C.E., Islam, A. K. M. S., Jones, R. G., Rahimi, M., Carrascal, D. R., Seneviratne, S. I., Servonnat, J., Sörensson, A. A., Sylla, M. B., Tebaldi, C., Wang, W. and Zaaboul, R. 2022. The Climatic Impact-Driver Framework for Assessment of Risk-Relevant Climate Information. *Earth's Future* **10**(11), e2022EF002803.
17. UK Climate Impacts Programme (UKCIP). 2011. Making progress: UKCIP & adaptation in the UK. Oxford, UK, UK Climate Impacts Programme.
18. UK Climate Resilience Programme (UKCR). 2019. Joint UKRI & Met Office Science Plan, UK Research and Innovation and Met Office.
19. Willows, R. I. and Connell, R. (Eds.) 2003 Climate adaptation: Risk, Uncertainty and Decisionmaking. Oxford, UK, UK Climate Impacts Programme.

Open Access This chapter is licensed under the terms of the Creative Commons Attribution 4.0 International License (http://creativecommons.org/licenses/by/4.0/), which permits use, sharing, adaptation, distribution and reproduction in any medium or format, as long as you give appropriate credit to the original author(s) and the source, provide a link to the Creative Commons license and indicate if changes were made.

The images or other third party material in this chapter are included in the chapter's Creative Commons license, unless indicated otherwise in a credit line to the material. If material is not included in the chapter's Creative Commons license and your intended use is not permitted by statutory regulation or exceeds the permitted use, you will need to obtain permission directly from the copyright holder.

CHAPTER 2

Climate Resilience: Interpretations of the Term and Implications for Practice

Kate Lonsdale, Nigel Arnell, Tim Coles, Kate Lock, Emer O'Connell, Paul O'Hare and Emma Tompkins

1 Introduction

The term 'resilience', which is integral to the UK Climate Resilience Programme (UKCR), has been used increasingly in academic, practice and public discourse around climate change, and crises more generally. The term's appeal comes from its ability to frame crises not as uncontrollable and uncertain phenomena to be feared, but as challenges over which one can triumph, with the potential for improving society. It has

---

Lead Author: Kate Lonsdale

Contributing Authors: Nigel Arnell, Tim Coles, Kate Lock, Emer O'Connell, Paul O'Hare & Emma Tompkins

---

K. Lonsdale (✉)
Climate Sense, Sheffield, UK
e-mail: kate.lonsdale@climatesense.global

N. Arnell
University of Reading, Reading, UK

© The Author(s) 2024
S. Dessai et al. (eds.), *Quantifying Climate Risk and Building Resilience in the UK*,
https://doi.org/10.1007/978-3-031-39729-5_2

an everyday meaning that emphasises interconnectedness and the 'bigger picture' of a system. Such optimistic and palatable qualities make it easy to see why it is popular. Who (or what) would not want to be resilient? [1].

However, the term is not universally liked. Some consider the concept too vague, 'restless' [2] or value-laden to be used in practice with any consistency. Some point to a tendency to focus on the technocratic features of resilience policy which provide limited potential to examine how power dynamics underpin how resilience is built ('resilience for whom?'). While this may act to promote political confidence, it does little to create the transformative change needed to unpick entrenched, structural inequalities in society. Framings emerging from recent academic interest in how resilience might better address power and agency present resilience not simply as response to a shock but rather as a dynamic capacity to be nurtured, developed, expanded and negotiated, given the right conditions at an individual, community, organisational or national scale [3, 4, 5, 6].

To provide some clarity, it is worth considering the key interpretations that are in use and demonstrated in the UKCR programme. We group the interpretations relevant to the work of UKCR into two categories: 'broad or narrow' and 'operational or place-based'. The aim of this is to draw out how the different framings influence the kind of activities needed to build greater resilience.

T. Coles
University of Exeter, Exeter, UK

K. Lock
University of Leeds, Leeds, UK

E. O'Connell
Greater London Authority, London, UK

P. O'Hare
Manchester Metropolitan University, Manchester, UK

E. Tompkins
University of Southampton, Southampton, UK

2 Broad or Narrow

The Intergovernmental Panel on Climate Change (IPCC) Sixth Assessment Report [7] glossary defines resilience as:

> The capacity of interconnected social, economic and ecological systems to cope with a hazardous event, trend or disturbance, responding or reorganising in ways that maintain their essential function, identity and structure.

However, we consider this too generalised to be useful in practice. Narrow framings have value where there is a specific service or entity that can be made resilient without requiring complex negotiation or inputs. The UK government's consultation on a National Resilience Strategy [8] defines resilience as:

> An ability to withstand and quickly recover from a difficult situation. This comes hand-in-hand with the idea of 'bouncing back', of returning to 'normal', of picking up where we left off before whatever difficulty or challenge we experienced.

Both 'bouncing back' and 'normal', written in inverted commas, highlight areas where there is controversy about what this means in practice. Is 'normal' simply the situation as it was before? Can a system (an asset, a community etc.) ever actually return to the same state after a crisis? Should the goal not be to 'build back better' and use the crisis as an opportunity to rethink and improve the existing situation? While this is a common definition of resilience and widely used in emergency planning, it represents a narrow interpretation, implying resilience is simply a *response* to external shocks to a system. Broader interpretations include concepts of anticipation and preparation to reduce exposure and vulnerability (e.g. through internal organisational processes), and see resilience depending not just on the nature of the external shocks but also on the factors that make the system of interest exposed and vulnerable. These different interpretations inevitably influence how organisations seek to enhance resilience.

3   Operational or Place-based

In addition to the distinction between broad or narrow resilience, further clarity of definition can be achieved by the distinguishing between 'operational resilience' or 'place-based resilience'.

Operational resilience refers to the resilience of a system, or component of a system, designed to deliver a specific outcome. It is used by operators of infrastructure or organisations responsible for delivering a service. For example, the Electronic Communications Resilience and Response Group (EC-RRG) [9] defines resilience as:

> The ability of an organisation, resource or structure to be resistant to a range of internal and external threats, to withstand the effects of a partial loss of capability and to recover and resume its provision of service with the minimum reasonable loss of performance.

Operational resilience is usually expressed in terms of technical standards of service or levels of protection for specific assets. Although this interpretation seems tightly and technically defined, there can still be problems with putting this into practice. As it is not feasible, nor cost effective, to make systems resilient to all conceivable shocks, decisions must be made about what is considered an acceptable standard of service. What 'probability of failure', or failure consequences, can we reasonably be expected to live with? Who defines what is a 'reasonable' loss of performance? A more laissez-faire approach would be for regulators to rely on 'best practice' and let organisations justify their own standards of service. This highlights the challenge of how to incorporate changing, uncertain conditions when setting standards for operational resilience in the face of climate change. For example, the Climate Change Committee [10] recommends adapting to a 2 °C world, assessing the risks for 4 °C and preparing for 'unpredictable extremes'—but how this can be done in practice needs to be defined carefully.

Place-based resilience relates resilience to a location rather than a specific system or service. High-level national strategies aspire to create 'resilient communities'. An example is Outcome 1 of Climate Ready Scotland Climate Adaptation Programme 2019–2024 [11], which is:

> Our communities are inclusive, empowered, resilient and safe in response to the changing climate.

The Environment Agency's National Flood and Coastal Erosion Risk Management (FCERM) strategy [12] meanwhile aims to achieve 'climate resilient places'. This could also apply to sectors, such as 'building a resilient health service'. Describing what place-based resilience looks like is challenging; places are exposed to multiple pressures, and people in a community will have diverse needs, expectations and levels of resource and motivation for engagement. What is considered 'resilient' for a locality, be it a neighbourhood, community, region or nation, is entirely socially constructed. Inevitably this involves political choices about priorities, where responsibility lies, and how decisions are made about what to fund. In the UKCR programme, this has been demonstrated in work developing principles of progressive resilience that recapture and recast the term in ways that resonate locally and with other drivers of change—such as reducing biodiversity loss, addressing poor mental health and access to green space—to make it more meaningful and applicable [13].

In recent years, a key critique of resilience has been that, despite appearing neutral and objective, resilience policies facilitate neoliberal shifts in responsibility for risk governance, particularly from the state to the private sector and communities. Such policies can be intensely competitive, creating both 'winners' and 'losers', and potentially maintaining an unequal status quo.

A distinction between operational and place-based resilience is helpful because approaches to characterising and measuring resilience are different, and because it influences discussions about how to achieve resilience. In summary, operational resilience is easier to characterise, measure and achieve than place-based resilience.

4 Implications for Building Resilience in the UK

How we interpret resilience is important. If we consider resilience as primarily dealing with external shocks, or as addressing underpinning features of the system that increase exposure and vulnerability, this clearly influences how we respond and the role and responsibilities of the state, communities and individuals. The Civil Contingencies Act https://www.legislation.gov.uk/ukpga/2004/36/contents (2004) commits to helping communities 'respond and recover' from external shocks rather than providing protection or support to address the reasons why some places or communities are more vulnerable. This operational interpretation of resilience focuses on measures to restore the status quo in the light of external events ('shocks'), rather than considering the underpinning characteristics of the system that create inequality. Place-based resilience, with its greater emphasis on anticipation and preparation to reduce exposure and vulnerability to subsequent loss (e.g. through land use planning, building regulations and social support), puts less emphasis on responding to external shocks. There is also a greater sense that 'bouncing back' to a previous state is not only unrealistic but also undesirable and unfair for many [4, 5]. The goal should thus be to 'bounce forward', or transform, to a fairer, better adapted state. To achieve this it helps to see resilience as something that is both contextual and negotiated by those with a stake in the outcome. This emphasises the need for effective mechanisms to facilitate difficult decisions about what is protected and what is lost, in the context of other drivers of change and limited resources.

5 What Next?

The term 'climate resilience' is likely to be with us for the near future. To be a useful concept for building climate resilience for the UK we need to ensure that the complexity of this can be addressed with the necessary level of detail and full consideration of interdependencies. To build climate resilience in the UK we need to frame resilience in a progressive way as 'bouncing forward', rather than back. This is an interpretation of resilience as a dynamic capacity that is negotiated and enabled, rather than a fixed state that is imposed. This requires:

- **Legitimate and inclusive mechanism(s) to engage across the whole of society.** We cannot save everything and must be selective. There will be co-benefits, unavoidable trade-offs and, inevitably, winners and losers. To achieve a just outcome this requires careful consideration, deliberation of 'resilience for whom?' and explicit discussion of winners and losers. Without careful inclusion of those most likely to 'lose', the voices of the vulnerable could be marginalised.
- **An informed public with access to accurate, salient information.** If negotiations about resilience need to happen at local, community and even sectoral scales, people need to be better informed of how they may be impacted by resilience policy and practice, directly and indirectly, through the places they live, how they travel, the food

they eat, the work they do, their personal connections and pursuits, and the public services they depend on.
- **An enabling environment with clear coordination and courageous leadership.** There is currently a lack of clear roles, responsibilities and accountability around building climate resilience in the UK. This includes mechanisms to raise public awareness, discuss what needs to change, to link the national policy processes to local experience and ensure it reduces inequality rather than exacerbates it. Citizens' assemblies and juries (such as the Rethinking Water Citizens' Juries https://consult.environment-agency.gov.uk/yorkshire/citizens-jury-for-the-river-wharfe-yorkshire-infor/#:~:text=The%20Rethinking%20Water%20Citizens'%20Jury%20was%20put%20together,experts%20on%20all%20aspects%20of%20the%20water%20environment) that use deliberative democracy to explore place-based approaches to resilience issues could be one solution, but most currently focus on achieving net zero. A greater emphasis on resilience—at least putting it on a par with net zero—is needed to stimulate this vital discussion and ensure fair and appropriate action.

References

1. White, I. and O'Hare, P. 2014. From Rhetoric to Reality: Which Resilience, Why Resilience, and Whose Resilience in Spatial Planning? *Environment and Planning C: Government and Policy* **32**(5), pp. 934–950.
2. Gleeson, B. 2008. Critical Commentary. Waking from the Dream: An Australian Perspective on Urban Resilience. *Urban Studies* **45**(13), pp. 2653–2668.
3. Eriksen, S.H., Nightingale, A.J. and Eakin, H. 2015. Reframing adaptation: The political nature of climate change adaptation. *Global Environmental Change* **35**, pp. 523–533.
4. Harris, L.M., Chu, E.K. and Ziervogel, G. 2017. Negotiated resilience. *Resilience: International policies, practices and discourses* **6**(3), pp. 196–214.
5. Ziervogel, G., Pelling, M., Cartwright, A., Chu, E., Deshpande, T., Harris, L., Hyams, K., Kaunda, J., Klaus, B., Michael, K., Pasquini, L., Pharoah, R., Rodina, L., Scott, D. and Zweig, P. 2017. Inserting rights and justice into urban resilience: a focus on everyday risk. *Environment and Urbanization* **29**(1), pp. 123–138.

6. McDonnell S. 2019. Other Dark Sides of Resilience: Politics and Power in Community-Based Efforts to Strengthen Resilience. *Anthropological Forum*.
7. Intergovernmental Panel on Climate Change (IPCC). 2022. *Climate Change 2022: Impacts, Adaptation, and Vulnerability. Contribution of Working Group II to the Sixth Assessment Report of the Intergovernmental Panel on Climate Change* [H.-O. Pörtner, D.C. Roberts, M. Tignor, E.S. Poloczanska, K. Mintenbeck, A. Alegría, M. Craig, S. Langsdorf, S. Löschke, V. Möller, A. Okem, B. Rama (eds.)]. Cambridge University Press. Cambridge University Press, Cambridge, UK and New York, NY, USA, pp. 3056.
8. HM Government. 2021. UK National Resilience Strategy: call for evidence, Cabinet Office.
9. Electronic Communications Resilience and Response Group. 2021. EC-RRG Resilience Guidelines for Providers of Critical National Telecommunications Guidelines. [Online]. Available at: https://www.gov.uk/guidance/electronic-communications-resilience-response-group-ec-rrg#ec-rrg-is-a-group-representing-all-elements-of-communications-services-in-order-to-promote-resilience-across-the-sector.
10. Climate Change Committee. 2021 Progress in adapting to climate change. 2021 Report to Parliament. [Online] Available at: https://www.theccc.org.uk/wp-content/uploads/2021/06/Progress-in-adapting-to-climate-change-2021-Report-to-Parliament.pdf.
11. Climate Ready Scotland: Climate Change Adaptation Programme 2019–2024. [Online] Available at: https://www.gov.scot/publications/climate-ready-scotland-second-scottish-climate-change-adaptation-programme-2019-2024/pages/4/.
12. Environment Agency: National Flood and Coastal Erosion Risk Management Strategy for England. [Online] Available at: https://www.gov.uk/government/publications/national-flood-and-coastal-erosion-risk-management-strategy-for-england--2.
13. Manchester Climate Change Agency. 2022 Manchester Climate Change Framework (2020–2025). 2022 update. [Online] Available at: https://www.manchesterclimate.com/sites/default/files/2022%20Update%20of%20the%20Manchester%20Climate%20Change%20Framework%20%282020-25%29%20AA.pdf.

Open Access This chapter is licensed under the terms of the Creative Commons Attribution 4.0 International License (http://creativecommons.org/licenses/by/4.0/), which permits use, sharing, adaptation, distribution and reproduction in any medium or format, as long as you give appropriate credit to the original author(s) and the source, provide a link to the Creative Commons license and indicate if changes were made.

The images or other third party material in this chapter are included in the chapter's Creative Commons license, unless indicated otherwise in a credit line to the material. If material is not included in the chapter's Creative Commons license and your intended use is not permitted by statutory regulation or exceeds the permitted use, you will need to obtain permission directly from the copyright holder.

PART I

Undertaking Resilience Research

CHAPTER 3

Towards a Step Change in Co-Production for Climate Resilience

Nicola Golding, Jenna Ashton, Kate Brown, Steven Chan, Tim Coles, Hayley Fowler, Elizabeth Fuller, Paula Harrison, Alice Harvey-Fishenden, Neil Macdonald and Christine Sefton

Abstract

- Co-production brought clear benefits to a range of projects across the UK Climate Resilience Programme (UKCR).
- Experiences were deeply context specific; dependent on those involved, their motivations and expectations.

Lead Author: Nicola Golding

Contributing Authors: Jenna Ashton, Kate Brown, Steven Chan, Tim Coles, Hayley Fowler, Elizabeth Fuller, Paula Harrison, Alice Harvey-Fishenden, Neil Macdonald & Christine Sefton

N. Golding (✉) · K. Brown · S. Chan · E. Fuller
Met Office, Exeter, UK
e-mail: nicola.golding@metoffice.gov.uk

© The Author(s) 2024
S. Dessai et al. (eds.), *Quantifying Climate Risk and Building Resilience in the UK*,
https://doi.org/10.1007/978-3-031-39729-5_3

- A range of barriers currently exist to achieving the benefits of co-production more fully.
- Skills associated with using co-productive approaches need to be developed, taught and mentored in the research community.

Keywords Co-production · Engagement · Community · Action research · Climate services

1 Introduction

This chapter records experiences of UKCR researchers whose projects incorporated co-production approaches to improve the usability, accessibility, relevance and credibility of outputs, and engage different groups of people in climate resilience. A workshop towards the end of the programme enabled researchers to reflect on benefits derived from this approach, and where and when barriers have existed and why. Participants outside academia, including sector experts and community participants involved in this research were invited, but were unable to attend. Different projects used co-production in a variety of ways, depending on the aims, motivations and theoretical backgrounds of those involved [1, 2].

J. Ashton
University of Manchester, Manchester, UK

S. Chan · H. Fowler
Newcastle University, Newcastle upon Tyne, UK

T. Coles
University of Exeter, Exeter, UK

P. Harrison
Centre for Ecology & Hydrology, Lancaster, UK

A. Harvey-Fishenden · N. Macdonald
University of Liverpool, Liverpool, UK

C. Sefton
University of Sheffield, Sheffield, UK

We outline key learnings and recommendations, while recognising that any learning is deeply context specific. We also highlight the need for a step change in fundamental aspects of research planning, funding, multi- and transdisciplinary working, to achieve the perceived benefits of co-production more fully.

2   What is Co-Production?

Co-production was popularised as a concept in the 1970s and has been taken up widely in the fields of public services, social care and health care. It challenged the knowledge-deficit model with a recognition that users of a service or product had valuable knowledge and experiences that could help to shape future research and development; it was becoming increasingly common in other sectors [3, 4, 5, 6]. No single definition of 'co-production' exists, but most reference 'equality of power', working together in 'partnership' or 'relationship' to generate knowledge or reach a 'collective goal' [7].

Recognition is growing that such collaborative approaches are needed to produce more usable and useful research and solutions to meet the challenges of societal resilience to a changing climate. Rapid growth in scientific understanding and technological capability has, to a large extent, outpaced the ability of scientists and other 'producers' to ensure outputs are relevant and tailored to society's needs. In addition, there is a recognised disconnect between the various disciplines involved in utilising climate information effectively in decision-making and adaptation, a lack of understanding of critical issues by decision-makers and a strong need for greater community engagement in action at the local level. Thus, co-production is increasingly being adopted in this field [8, 9, 10]. Various research initiatives have provided useful guidance and principles for co-production, particularly within climate services [11, 12, 13, 14, 15]; this has resulted in a shift away from the often unhelpful binary framing of 'producer' and 'user', towards a recognition that many different stakeholders hold valuable information and knowledge, resulting in an improved power-balance that can contribute to decision-making for resilient societies. However, there is also an increasing recognition that co-production must be done appropriately, with a shared understanding of what is expected, and if not given due consideration this can cause damage and a strong disengagement. UKCR projects have contributed further to this literature, through combined learning of what works, and what the remaining challenges are for climate resilience co-production.

3 What Works Well

3.1 Gathering Community Experience

The 'Creative Climate Resilience' project has demonstrated success in using place-based folks arts and socially engaged practices—for residents, policymakers and local authority workers—to co-produce knowledge as part of a community development and social change framework [16]. This has supported participants working together to firstly identify their own needs and actions for local climate planning; secondly take collective action, identifying and using their strengths and resources; thirdly develop confidence, skills and knowledge for mitigation and adaptation; fourthly challenge unequal power relationships; and finally promote social justice, equality and inclusion.

Throughout the project, there were numerous encounters with participants through arts-based research methods, including place exploration, visual arts workshops, heritage interpretation, recorded interviews and stories, animations, puppetry artist residency, creative writing and song-writing—alongside geographic information system (GIS) spatial analysis and biodiversity data analysis. These processes built community capacity, connectivity and skills, and drew out local knowledge. By working in this way, strong relationships were built between residents, local authority staff and organisations and the research team. This built trust in the process and contributed directly to local authority neighbourhood climate action planning and wider legacy work in community development and social action, such as contributing findings around resilience to local authority decision-makers; supporting resident-led fundraising and capacity building; informing landscape decision-making; enabling resident self-expression for communicating needs and opportunities; encouraging political literacy; and further research and development for establishing new community assets.

3.2 Sustaining Engagement Throughout

The original Met Office City Pack, developed through the 'Meeting Urban User Needs' project, is a successful climate service prototype co-produced in close partnership with Bristol City Council. This project highlighted the advantages of joint initiation—and an opening discussion centred on what would add value—before defining outputs or outcomes. Regular workshops, online interaction and iteration of prototypes throughout the process created a trusting relationship between the local authority and researchers, resulting in a highly bespoke service for the city of Bristol, which is deemed successful by a range of stakeholders.

Another project, 'CLandage', saw landscapes and cultural heritage researchers partner with Tasglann nan Eilean Siar (the Hebridean Archives), Staffordshire Record Office and Historic England, to capture individual and community experiences of storms, floods and droughts, and how they have adapted and developed resilience through time [17]. Partners were involved from project inception, with relationships already fostered through previous research. Prior experience of working together proved invaluable, particularly as the project evolved during the COVID-19 pandemic, with the loss of face-to-face activities, travel restrictions and closure of archives and museums. These altered working practices actually resulted in a closer partnership within 'CLandage', and more effective co-production.

The 'Creative Climate Resilience' project also demonstrated the need for co-production across a diverse network of individuals and groups, and at different moments and intensities during the project. This network included the interdisciplinary research team, residents, local authority civil servants and neighbourhood teams, local environment organisations, community development organisations and artists.

3.3 Getting Creative with Storytelling

Various projects advocated the use of storytelling to facilitate co-production. In 'CLandage', online poetry workshops led by a local poet—working with archive materials, supplied by Staffordshire Record Office—encouraged participants to reflect on local experiences of flooding through poetry [18]. Also, a creative-maker led small workshops, using old pictures and reports of drought, to stimulate and explore memories of flooding and drought. In addition, a storyteller led a series of walking tours, collecting memories and exploring ideas of flooding through traditional stories and oral histories. In each instance, the creative approaches were initiated by Staffordshire Record Office and led by individuals from the local area, placing high value and emphasis on experiential knowledge, thereby enabling and empowering communities.

3.4 Balancing Power and Managing Expectations

The co-productive approach taken by the 'Creative Climate Resilience' project helped to navigate differences in income, health and well-being, and education. It also succeeded in overcoming entrenched apathy with local political systems, and a sense of disenfranchisement with local decision-making. The project was part of a much longer process of supporting community development with climate change challenges. Researchers connected into and supported existing community practices, such as resident-led networks, local charity initiatives and community spaces, acknowledging the wider societal action independent of the research project. This approach, alongside a continuous physical and emotional presence in the area as part of the socially engaged arts methodology, and flexibility to respond to challenges and opportunities as they arose, led to a more equitable balance of power and constructive conversation.

The 'UK-SSPs' project created UK and nation-specific socioeconomic scenarios for use alongside climate change projections, to assess risk, vulnerability and resilience. The need to ensure consistency with previous global SSPs limited the extent to which co-production was possible and required careful management of expectations. To minimise constraints on stakeholder imagination, stakeholders were asked to identify and cluster socioeconomic drivers that they considered particularly important (and uncertain) before being introduced to the global SSPs, onto which they could then map the UK-specific drivers. The process was highly iterative, balancing the need for consistency and legitimacy with stakeholder creativity, in order to develop a set of UK-specific SSPs that are locally comprehensive and relevant, yet consistent with global SSPs [19, 20].

The project 'Transport/Energy Climate Services' also found it beneficial to proactively manage expectations of the co-production process. User expectation was often high, with some expecting "the ideal solution" after a relatively light investment of resource and engagement; in reality, what emerged from the project was simply the first step in solving the problem. This approach also recognises the different expectations and approaches to co-production across disciplines, as well as the need for transparency and agreement.

3.5 Experimenting with Upscaling

The City Pack resource (outlined above) was subsequently rolled out to other UK cities, but with less co-production; while city-specific information changed, the template remained the same and user engagement was, therefore, less-intensive. Researchers explored different approaches to upscaling co-production, and assessed value lost if co-production was less central. Their experience suggests that upscaling a climate service, which has been co-produced with one set of stakeholders, can provide a useful and usable service for others.

4 Emerging Challenges and Opportunities

The UKCR experience suggests that, when done well, co-production is an effective way to bring climate information together with other forms of knowledge to support resilience building at community, organisational and policy levels. The remainder of this chapter offers a set of recommendations emerging from the UKCR programme on how to set up effective co-productive approaches.

4.1 Focus on the Process, Not Just Outputs

One of the main barriers to successful co-production among projects was where the research process was fixed on pre-defined outputs. Projects that saw the primary aim of co-production as discovery, or knowledge sharing, were more open to learning and bringing in others' perspectives. Linked to this was a need to be flexible, to allow the detail of what is considered a 'useful output' to emerge through the process of engagement. Where some pre-definition of outputs may be required for funding purposes, flexibility is needed to allow for changes as a wider understanding of the nature of the problem deepens and assumptions are challenged. Not fixing too early on an idea avoids creating outputs that later turn out to be unfit for purpose or missing the point. Participants particularly valued the "opportunities for serendipity" afforded by the flexibility of co-production, quoting:

> The story changes throughout the process... this is fundamental to co-production. Everything is part of the process

The more co-creative projects within UKCR enabled this process, as their goal was to learn and share, rather than to produce specific outputs. The experience and maturity of researchers and their relationship with partners were seen as another factor in how effective projects were in setting up more flexible research approaches. Such researchers were better able to craft research proposals that satisfied funding requirements, while also setting up participatory processes with sufficient flexibility to allow for emergent processes. To achieve this, there is a need to develop the skills necessary to draft research proposals with a 'design for learning' approach, as well as a shift in the way research is funded to focus on goals and outcomes rather than specific outputs.

Researchers from the 'MAGIC' project reflected on the element of the unknown, emphasising the need to have shared understanding of aspirations and expectations from the early stages of the research process. This was shared by 'Risky Cities', which gradually increased the use of immersive co-production until "neither party had control" over what the other party was going to do. In this case, co-production was initiated by the National Youth Theatre, who approached scientists to talk to them about the climate crisis; it soon became apparent that both parties had a lot more to offer each other. This learning, and the fundamental need to address power inequalities in co-production, is reflected in the wider literature [9].

4.2 Revise Funding Structures and Timescales

Many of the challenges relate to limitations of current funding structures and timescales. Some researchers argued that all participants in the process (not just researchers) need to be funded, as without this non-academic partners in the research found it hard to make time to participate or lacked the incentive to do so. However, some researchers felt that co-production was most successful when stakeholders were motivated to engage without funding.

There is also a need to move away from rigid project timescales to do justice to co-production and allow it to be truly emergent, because co-production requires time to build relationships and a culture of trust to enable the work to flourish. As such, traditional finite-length project funding can be limiting. Flexibility in the funding model offers opportunities to shape and respond to emergent processes, and longer-term funding furthermore enables breaking down of disciplinary boundaries. Researchers engaged in co-developing prototype climate services for the energy sector revealed that "There was a lot of early engagement, but the needs were so diverse it took a long time to identify where to work together". It was also reportedly difficult firstly to establish what was critical and urgent for users, and secondly to manage expectations in terms of progress towards the ideal solution.

Many researchers attributed their success to pre-existing relationships; in many cases, links were developed through previous projects, and so a co-production process emerged organically benefitting significantly from the existing understanding and trust [19, 20].

4.3 Promote New Measures of Success

There is a need to recognise the importance of relational, embodied learning, connection, and sharing knowledge, fun and trust, alongside more traditional outputs, such as peer-reviewed papers and tangible products. New measures of success are needed to evaluate the process, in addition to the outcome, and to reflect different motivations and requirements of participants [21, 22]. Within projects there was a general recognition of the value of the process (e.g. building trust, collective decision-making) but also a perceived challenge that these are not generally recognised by funders. Researchers commented that there is a need to "understand other ways of doing knowledge", and that funding needs to be "targeted towards processes not just outputs", requiring a change in mindset.

One barrier to co-production and associated transdisciplinary research was the perceived lack of incentive within an academic career. Promotion criteria at universities are, in many cases, still skewed towards the disciplinary academic; better recognition of transdisciplinary processes, and weighting them for promotion criteria, is needed. This experience can be particularly acute for early career academics attempting to acquire permanent academic posts, where inter- and transdisciplinary skill sets may be perceived as lacking a clear focus. However, the practice of transdisciplinary research had a positive impact on more experienced researchers, through advancing networks and opportunities for future impactful research. Public scholarship and other motivations for co-production were seen as critical to its success, while encouragement for early career researchers is needed to underpin shifting attitudes.

4.4 Invest in Multidisciplinary Approaches

Successful co-production for climate resilience demands input from a range of disciplines and stakeholders, including researchers, practitioners and action-takers [23, 24]. Researchers commented that key skills were missing from project teams, as "academics can only help with part of the problem". 'MAGIC' addressed this by implementing a community-led approach to reducing flood risk, achieved through a case study of the flood-vulnerable region around Hull. Hull and East Riding Timebank, one of the key partners, provided a range of expertise and skills beyond those traditionally included in projects.

Researchers also argued that funding mechanisms should be available to resource additional expertise in response to specific issues as they emerge. However, the transient nature of this solution provides its own challenges; therefore, more enduring teams, capturing the requisite skills, expertise and competencies are also needed.

5 Conclusions

The experiences of UKCR researchers have confirmed a range of benefits of co-production. In addition, the challenges cited highlight the need for a step change in co-production for climate resilience (summarised in Fig. 1). Current approaches to research design, planning and funding, as well as the skills of researchers, can present obstacles to fully achieving the perceived benefits of co-production. One aspect that all projects reflected on is the fundamental role of the 'quality of relationships'. In many cases, co-production success was a function of the relationship; a lack of time or close engagement contributed to power inequalities, which if not addressed could inadvertently cause harm or derail knowledge creation into predetermined science-based frameworks. Several other key considerations also emerged:

- Engaging suitable contacts at the appropriate level of an organisation is critical. Contacts need to be sufficiently well connected, but with time and inclination to engage closely and sustain the relationship.

- Having a flexible approach and realistic expectations are crucial to success. The ability to build relationships and establish trust are important skills, which should be practised, taught and cultivated.
- Prior experience of co-production can engender the confidence needed to take perceived risks or develop flexible research proposals. It can also equip the researcher with the confidence to recognise and make changes when a partnership is not effective.

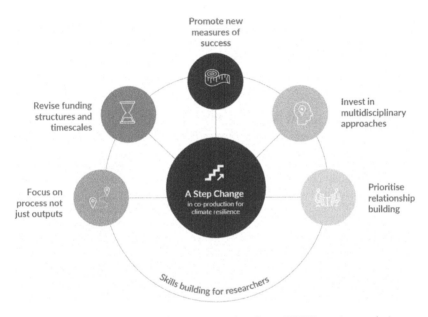

Fig. 1 Summary of the key themes emerging from UKCR projects relating to a required step change in co-production research for climate resilience

The need for relationship building was accentuated by the COVID-19 pandemic restrictions. Close relationships allowed for a more agile response to changing circumstances. This is another argument for investing in such skills—and integrating time for nurturing relationships into career development and project design.

Finally, it is clear from the UKCR programme that there are a range of effective and transformative approaches to co-production, driven by different disciplinary backgrounds, motivations and expectations. Many of the most successful examples enacted co-production continuously throughout the research, with creation of knowledge and understanding the primary aim, and other pre-defined outputs secondary to this. In many cases, the aspiration to co-produce was juxtaposed with reality and practicality, particularly in the context of restrictions during the COVID-19 pandemic. There is a spectrum of approaches, and each has its relative strengths. Notwithstanding, in strong agreement with previous literature [25, 26, 27], there is a need for greater transparency and shared expectation regarding co-production: what it means in each context, how it will be achieved, and what the anticipated benefits are for all participants.

References

1. Miller, C.A. and Wyborn, C. 2020. Co-production in global sustainability: Histories and theories. *Environmental Science and Policy* **113**(2020), pp. 88–95.
2. Carter, S., Steynor, A., Vincent, K., Visman, E. and Waagsaether, K. 2019. Co-production of African weather and climate services. Manual, Cape Town: Future Climate for Africa and Weather and Climate Information Services for Africa [Online] Available at: https://futureclimateafrica.org/coproduction-manual.
3. Vargo, S. and Lusch, R. 2004. Evolving to A New Dominant Logic for Marketing. *Journal of Marketing* **68**, pp. 1–17.
4. Auh, S., Bell, S. J., McLeod, C. S. and Shih, E. 2007. Co-Production and Customer Loyalty in Financial Service. *Journal of Retailing* **83**(3), pp. 359–370.
5. Lember, V., Brandsen, T. and Tõnurist, P. 2019. The potential impacts of digital technologies on co-production and co-creation. *Public Management Review* **21**(11), pp. 1665–1686.
6. Solman, H., Smits, M.,van Vliet, B., Bush, S. 2021. Co-production in the wind energy sector: A systematic literature review of public engagement beyond invited stakeholder participation. *Energy Research and Social Science* **72**, 101876.
7. Coutts, P. 2019. The many shades of co-produced evidence. Carnegie UK Trust Publication.

8. Kruk, M. C., Parker, B., Marra, J. J., Werner, K., Heim, R., Vose, R. and Malsale, P. 2017. Engaging with users of climate information and the co-production of knowledge. *Weather Climate and Societ*, **9**, pp. 839.
9. Bremer, S., Wardekker, A., Dessai, S., Sobolowski, S., Slaattelid, R. and van der Sluijs, J. 2019. Toward a multi-faceted conception of co-production of climate services. *Climate Services* **13**, 2019, pp. 42–50.
10. Turnhout, E., Metze, T., Wyborn, C., Klenke, N. and Louder, E. 2020. The Politics of Co-Production: Participation, Power, and Transformation. *Current opinion in environmental sustainability* **42**, pp. 15–21.
11. Vincent, K., Daly, M., Scannell, C. and Leathes, B. 2018. What can climate services learn from theory and practice of co-production? *Climate Services* **12**, pp. 48–58.
12. Visman, E., Audia, A., Crowley, F., Pelling, M., Seigneret, A. and Bogosyan, T. 2018. Underpinning principles and ways of working that enable co-production: Reviewing the role of research. BRACED Learning Paper #7, King's College London/BRACED.
13. O'Connor, R. A., Nel, J.L., Roux, D.J., Lim-Camacho, L., van Kerkhoff, L. and Leach, J. 2019. Principles for evaluating knowledge co-production in natural resource management: Incorporating decision-maker values. *Journal of Environmental Management* **249**.
14. Norström, A. V., Cvitanovic, C., Löf, M.F., West, S., Wyborn, C., Balvanera, P., Bednarek, A.T., Bennett, E.M., Biggs, R., de Bremond, A., Campbell, B.M., Canadell, J.G., Carpenter, S.R., Folke, C., Fulton, E.A., Gaffney, O., Gelcich, S., Jouffray, J.B., Leach, M., Le Tissier, M., Martín-López, B., Louder, E., Loutre, M.F., Meadow, A.M., Nagendra, H., Payne, D., Peterson, G.D., Reyers, B., Scholes, R., Speranza, C.I., Spierenburg, M., Stafford-Smith, M., Tengö, M., van der He, S. l, van Putten, I. and Österblom, H. 2020. Principles for knowledge co-production in sustainability research. *Nature Sustainability*, pp. 1–9.
15. Máñez Costa, M., Oen, A.M.P., Neset, T-S., Celliers, L., Suhari, M., Huang-Lachmann, J-T., Pimentel, R., Blair, B., Jeuring, J., Rodriguez-Camino, E., Photiadou, C., Columbié, Y.J., Gao, C., Tudose, N.-C., Cheval, S., Votsis, A., West, J., Lee, K., Shaffrey, L.C., Auer, C., Hoff, H., Menke, I., Walton, P. and Schuck-Zöller, S. 2021. Co-production of Climate Services. CSPR Report No 2021:2, Centre for Climate Science and Policy Research, Norrköping, Sweden.
16. Banks, S. and Westoby, P. 2019. Ethics, equity and community development, Bristol, Policy Press.
17. Naylor, S., Macdonald, N., Bowen, J. P. and Endfield, G. 2022. Extreme weather, school logbooks and social vulnerability: The Outer Hebrides, Scotland, in the late nineteenth and early twentieth centuries. *Journal of Historical Geography* **78**, pp. 84–94.

18. Wardle Woodend, M., Harvey-Fishenden, A. and Macdonald, N. (Eds.) 2022. Flood and Drought Poetry: Experiences of Weather Extremes in Staffordshire. Staffordshire: Dreamwell Writing Limited. [Online] Available at: http://www.dreamwellwriting.simplesite.com/.
19. Pedde, S., Harrison, P.A., Holman, I.P., Powney, G.D., Lofts, S., Schmucki, R., Gramberger, M. and Bullock, J.M. 2020. Enriching the Shared Socioeconomic Pathways to co-create consistent multi-sector scenarios for the UK. *Science of the Total Environment* **756**.
20. Harmáčková, Z.V., Pedde, S., Bullock, J.M., Dellaccio, O., Dicks, J., Linney, G., Merkle, M., Rounsevell, M.D.A., Stenning, J. and Harrison, P.A. 2022. Improving regional applicability of the UK Shared Socioeconomic Pathways through iterative participatory co-design. *Climate Risk Management*, **37**, pp. 100452.
21. Arnott, J.C., Kirchhoff, C.J., Meyer, R.M., Meadow, A.M. and Bednarek, A.T. 2020. Sponsoring actionable science: what public science funders can do to advance sustainability and the social contract for science. *Current Opinion in Environmental Sustainability* **42**, pp. 38–44.
22. Wall, T.U., Meadow, A.M. and Horganic, A. 2017. Developing Evaluation Indicators to Improve the Process of Coproducing Usable Climate Science. *Weather, Climate, and Society* **9**, pp. 95–107.
23. Polk, M. 2015. Transdisciplinary co-production: Designing and testing a transdisciplinary research framework for societal problem solving. *Futures* **65**, pp. 110–122.
24. Steynor, A., Lee, J. and Davison, A. 2020. Transdisciplinary co-production of climate services: a focus on process. *Social Dynamics* **46**(3), pp. 414–433.
25. Bremer, S. and Meisch, S. 2017. Co-production in climate change research: reviewing different perspectives. *WIREs Climate Change* **8**, pp. e482.
26. Lemos, M.C., Arnott, J.C., Ardoin, N.M., Baja, K., Bednarek, A.T., Dewulf, A., Fieseler, C., Goodrich, K.A., Jagannathan, K., Klenk, N., Mach, K.J., Meadow, A.M., Meyer, R., Moss, R., Nichols, L., Sjostrom, K.D., Stults, M., Turnhout, E., Vaughan, C., Wong-Parodi and G., Wyborn, C. 2018. To co-produce or not to co-produce. *Nature Sustainability* **1**, pp.722–724.
27. Meadow, A.M., Ferguson, D.B., Guido, Z., Horangic, A., Owen, G. and Wall, T. 2015. Moving toward the Deliberate Coproduction of Climate Science Knowledge. *Weather, Climate, and Society* **7**, pp. 179–191.

Open Access This chapter is licensed under the terms of the Creative Commons Attribution 4.0 International License (http://creativecommons.org/licenses/by/4.0/), which permits use, sharing, adaptation, distribution and reproduction in any medium or format, as long as you give appropriate credit to the original author(s) and the source, provide a link to the Creative Commons license and indicate if changes were made.

The images or other third party material in this chapter are included in the chapter's Creative Commons license, unless indicated otherwise in a credit line to the material. If material is not included in the chapter's Creative Commons license and your intended use is not permitted by statutory regulation or exceeds the permitted use, you will need to obtain permission directly from the copyright holder.

CHAPTER 4

Learning from Organisational Embedding for Climate Resilience

Kate Lonsdale, Tim Coles, Paul O'Hare, Caitlin Douglas, Stephen Scott-Bottoms, Alan Kennedy-Asser, Charles Rougé and Corinna Wagner

Abstract

- This paper describes the UK Climate Resilience Programme (UKCR) Embedded Researcher (ER) scheme, in which 13 researchers were 'embedded' within 'host' organisations to undertake a research project of mutual interest.

Lead Author: Kate Lonsdale

Contributing Authors: Tim Coles, Paul O'Hare, Caitlin Douglas, Stephen Scott-Bottoms, Alan Kennedy-Asser, Charles Rougé & Corinna Wagner

Findings are based on conversations with all UKCR embedded researchers and several hosts.

K. Lonsdale (✉)
Climate Sense, Sheffield, UK
e-mail: kate.lonsdale@climatesense.global

© The Author(s) 2024
S. Dessai et al. (eds.), *Quantifying Climate Risk and Building Resilience in the UK*,
https://doi.org/10.1007/978-3-031-39729-5_4

- There was considerable interest in the scheme from public, private and third sector organisations.
- The COVID-19 lockdowns limited the extent that ERs could physically work within their host organisation, but embedding and collaborative working was still achieved.
- ERs and hosts agreed that the approach enabled more fit-for-purpose outcomes than through traditional research or consultancy; future schemes could include 'host' staff spending time in research institutions to better understand the nature of academic knowledge production.
- Factors influencing effectiveness included the perception of being 'on the inside' of the organisation; the flexibility of the research workplan; the openness of the ER and host to learning; a facilitative and curious outlook; and the commitment to achieve mutually beneficial goals.

Keywords Embedding · Transdisciplinarity · Resilience · Research · Co-production

T. Coles · C. Wagner
University of Exeter, Exeter, UK

P. O'Hare
Manchester Metropolitan University, Manchester, UK

C. Douglas
Climate Change Committee, London, UK

S. Scott-Bottoms
University of Manchester, Manchester, UK

A. Kennedy-Asser
University of Bristol, Bristol, UK

C. Rougé
University of Sheffield, Sheffield, UK

1 INTRODUCTION

As the imperative for effective responses to our changing climate grows, so too do calls for more agile ways to bring climate-related information into decision-making, and allowing policy and practice to inform research directions and approaches [1, 2]. However, this is not straightforward—research must engage with a diversity of stakeholders and sectors and in a range of organisational settings. Spanning science, policy and practice requires careful brokering, convening and sense-making to ensure climate information fits organisational contexts, and reciprocally shapes ongoing production of knowledge [3–6].

Traditional climate science communication is often portrayed as the linear transmission of information and knowledge between 'producers' and 'users' [7]. In this chapter, we present the UKCR's Embedded Researcher (ER) scheme as an alternative and reflect on its relative merits and achievements. By placing researchers within host organisations, the scheme acknowledges the importance of organisational contexts, with climate science as one of several sources of decision-relevant information needed. Novelty and research excellence emerge from shared development and discovery, ensuring information is 'actionable' and fit-for-use. An underlying premise is that more trusting researcher-host relationships evolve through immersion, as well as close, collaborative working towards a common purpose. This supports deeper exploration of constraints to action, resulting in significant and meaningful outcomes.

2 THE EMBEDDED RESEARCHER SCHEME

The scheme comprised two cohorts of researchers embedded for up to 12 months. Cohort 1 had a two-stage application process: following a call for hosts to propose research questions, UK Research and Innovation (UKRI)/Natural Environment Research Council (NERC) published the 20+ research ideas, inviting interested researchers to contact the host organisations to prepare a collaborative bid. Cohort 2 had a single-stage application process, whereby researchers were invited to develop projects directly in partnership with host organisations.

2.1 Funded Projects and Outcomes

The scheme had an enthusiastic response from 35 public, private and third sector organisations. Table 1 summarises the 13 projects funded, and the range of contexts and outcomes achieved. Feedback from hosts was positive, but inevitably some projects were more successful and collaborative than others. The factors that influenced this are addressed later in this paper.

2.2 A Note on Embedding During a Global Pandemic

Cohort 1 started mid-pandemic (COVID-19). Through online working, attending meetings and sharing data was easy, but virtual working posed limits to the extent ERs could be fully embedded. For Cohort 2, a hybrid way of working was possible, allowing face-to-face meetings and events during spring and summer 2022. During the period of the scheme, online or hybrid working and 'only travel if you have to' advice became normal, inevitably changing expectations of embedding.

3 How did researchers and hosts experience the ER scheme?

Table 1 demonstrates the variety of organisational settings and projects under the UKCR ER scheme. The ER experience also varied, in terms of how embedded the researchers felt and how collaborative the work was. Remote working during the COVID-19 lockdown reduced opportunities for informal contact with colleagues, which made it harder for some to distinguish their work from consultancy or traditional research. Others reported feeling well-embedded, and that 'moving online' made joining meetings with senior colleagues and outside organisations easier.

In terms of added value, the ERs increased the capacity, impact and reach of hosts' work, contributed new ideas and ways of working, and were able to take advantage of opportunities that arose, such as the drafting of the adaptation and resilience section of Manchester's climate policy https://www.manchesterclimate.com/framework-2020-25. Some elements are indistinguishable from what could be achieved through traditional research or consultancy, such as giving advice on how to use probabilistic projections or providing additional 'bandwidth' to support ongoing work. The following quotes captured from hosts and ERs and

Table 1 The details of the funded projects across the two cohorts

Host	Academic institution	Project title and outcome	Duration (intended)	Academic status at start
COHORT 1: (September 2020–December 2021)				
Space4 Climate[a] with London Climate Change Partnership[b]	King's College London[c]	**Climate Stress Testing** *Outcome:* Brought together stakeholders in the UK food supply chain and the earth observation industry to improve the UK's food security	12 months 0.6 FTE (later 0.4 FTE)	Research Associate
Environment Agency[d]	Newcastle University[e]	**Environment Agency Incident Response** *Outcome:* Clearer characterisation and quantification of current Environment Agency flood and drought incident response activity, and capacity required for future climates	12 months 0.5 FTE	Lecturer
Department for Education[f]	University College London[g]	**ARID** *Outcome:* Enhanced characterisation, quantification and communication of climate-related school building asset management risks through developing adaptation pathways to rising heat stress	12 months Full time	Research Assistant (Completing PhD)
Manchester Climate Change Agency[h]	Manchester Metropolitan University[i]	**Manchester Climate Action** *Outcome:* Established a baseline assessment of Manchester's climate risk, and a policy and action-planning framework to enable Manchester to adapt to and increase resilience to climate variability	12 months 0.8 FTE	Senior Lecturer
Bristol City Council[j]	University of Manchester[k]	**Bristol Heat Resilience** *Outcome:* Co-developed a Heat Vulnerability Index and a Heat Resilience Plan for Bristol to support the City Council in developing heat risk reduction strategies and increased resilience for citizens, communities and businesses[l]	17 months 1.0 FTE (with additional funding from host)	Researcher (Completing PhD)

(continued)

Table 1 (continued)

Host	Academic institution	Project title and outcome	Duration (intended)	Academic status at start
Anglian Water[m]	University of Sheffield[n]	**Water Sector Resilience** *Outcome:* Initiated a long-term collaboration to identify and address gaps in climate adaptation in water resource systems to support better system-level adaptation planning	12 months 0.5 FTE	Lecturer
COHORT 2: November 2021–October 2022 (and ongoing at time of writing)				
Leeds City Council[o] with Yorkshire and Humber Climate Commission[p]	University of Manchester[q]	**Yorkshire Climate Action** *Desired outcome:* Clarity on responsibility for implementing the Yorkshire and Humber Climate Commission's new Climate Action Plan in Leeds City Council, using performance to kick-start conversations with different service areas	12 months 0.2 FTE	Professor
City of London Corporation[r]	British Geological Survey[s]	**London Climate Action** *Desired outcome:* Improved understanding of how urban subsurface space can be used to deliver the City of London's Climate Action Strategy and improve climate resilience	12 months 0.5 FTE	Researcher
JBA Consulting[t]	Newcastle University[u]	**Stochastic Simulation** *Desired outcome:* Improved understanding and use of stochastic weather generators in applied UK climate resilience projects, with a focus on flood and water management	12 months 0.75 FTE	Research Associate
Climate Northern Ireland[v]	University of Bristol[w]	**Once Upon a Time** *Desired outcome:* Improved two-way dialogue between rural/agriculture and academic/policy communities, leading to better understanding of climate risk and resilience options[x]	12 months 0.55 FTE	Research Associate

(continued)

Table 1 (continued)

Host	Academic institution	Project title and outcome	Duration (intended)	Academic status at start
Time and Tide Bell[y]	University of Exeter[z]	**Time and Tide** Desired outcome: Greater understanding of how Time and Tide Bells, specifically, and science-informed art more generally help communities become more resilient in the face of climate change and socioeconomic inequalities	12 months 0.2 FTE	Professor
Church of England[aa]	University of Manchester[ab]	**Resilience for Churches** Desired outcome: Enhanced community climate resilience and protection of Church of England's churches and other heritage buildings through the collation and dissemination of successful climate adaptation strategies already in use	12 months 0.8 FTE	Researcher (Completing PhD)
National Trust[ac] and Historic Environment Scotland[ad]	University of Exeter[ae]	**Tourism Adaptation** Desired outcome: Greater awareness of the potential impact of future climate change and scenarios on visitor business	12 months 0.4 FTE	Professor

Notes [a]https://space4climate.com/; [b]https://climatelondon.org/; [c]https://www.kcl.ac.uk/; [d]https://www.gov.uk/government/organisations/environment-agency; [e]https://www.ncl.ac.uk/; [f]https://www.gov.uk/government/organisations/department-for-education; [g]https://www.ucl.ac.uk/; [h]https://www.manchesterclimate.com/; [i]https://www.mmu.ac.uk/; [j]https://www.bristol.gov.uk/; [k]https://www.manchester.ac.uk/; [l]https://www.bristol.gov.uk/council-and-mayor/policies-plans-and-strategies/energy-and-environment/the-keep-bristol-cool-mapping-tool; [m]https://www.anglianwater.co.uk/; [n]https://www.sheffield.ac.uk/; [o]https://www.leeds.gov.uk/; [p]https://yorksandhumberclimate.org.uk/; [q]https://www.manchester.ac.uk/; [r]https://www.cityoflondon.gov.uk/; [s]https://www.bgs.ac.uk/; [t]https://www.jbaconsulting.com/; [u]https://www.ncl.ac.uk/; [v]https://climatenorthernireland.org.uk/; [w]https://www.bristol.ac.uk/; [x]https://ukcrp.shinyapps.io/AgricultureNI/; [y]https://timeandtidebell.org/; [z]https://www.exeter.ac.uk/; [aa]https://www.churchofengland.org/; [ab]https://www.manchester.ac.uk/; [ac]https://www.nationaltrust.org.uk/; [ad]https://www.historicenvironment.scot/; [ae]https://www.exeter.ac.uk/

hosts highlighted what made the UKCR ER approach unique—and thereby more impactful:

- Researchers gained a deeper understanding of the organisation.

Embedding gave me a more complete understanding of how the organisation works and what information they need to make decisions (e.g. sectors, geographic regions, level of detail). (ER)

All of the materials that he's [the ER] developing for us – and the workshops he's going to run – suit us because he's embedded enough to understand how to do them in a way that works in our slightly odd and complicated organisation. (Host)

- Through the extended collaboration, there was time to experiment and to revise.

Through exploring something together, we gained a first step in a new area with an array of challenges we need to overcome to move forward. This was quite a high-risk piece of work which would not have worked as a consultancy project. (Host)

- Researchers and hosts forged long-term relationships.

I had time to work with the host to identify operational gaps for future research. We identified two such gaps, leading to an Engineering and Physical Sciences Research Council (EPSRC) grant and an Industrial Cooperative Awards in Science & Technology (CASE) doctoral studentship. (ER)

This was a great way to establish a working relationship. I hope we will have many more collaborations with our ER. Some papers are planned and the conversation is certainly ongoing. (Host)

- Researchers were able to act as a catalyst.

This allows you to join up a whole bunch of people who are thinking, doing, or wanting to do things in their community. (ER)

- Researchers were able to develop a more meaningful dialogue with hosts.

Embedding is the way to get to what people want, which is action. Because you're in the inside there is less of a perceived barrier. You have more time to get feedback and think through what will work. (ER)

By listening to people, their identities, their concerns, and feeling part of it yourself, you want to avoid solutions that don't work, like a policy prescribed by someone who has no idea about that locality, or the people's lives in that locality. (ER)

- The scheme supported more effective communication and engagement.

I used to do a lot of public engagement. Embedding is just a totally different way of doing it - a better way. Now just getting up on a stage and giving a public talk seems like a waste of time. This is a much deeper way of engaging because you're part of it. (ER)

I have changed how I communicate. It used to be very one-directional. Good storytelling is patient, slow and takes time. And that's what embedding does. It is not standing up and giving a talk and answering questions at the end, and then going home. It is weeks of having conversations and eventually you get these 'aha!' moments. (ER)

- Links were forged across boundaries and between internal/external expertise and action.

We talk about interdisciplinarity, but this is beyond that. You're working across disciplines, and you're still doing researcher stuff, but you're also a facilitator, a convenor, a planner. And you're responding to whoever you're working with. (ER)

- Researchers gained new perspectives on the knowledge needed for decision-making.

It's a humbling experience, the recognition that there is whole set of other types of knowledge, experience and wisdom out there. (ER)

I used to think there was very little information on impacts in Northern Ireland. Now, after a year of meeting lots of people, actually there's so much knowledge out there. (ER)

I'm not as quick to go down a scientific rabbit hole and lose sight of 'what do we actually do to solve this?'. I've realised that often tiny, tiny details in the scientific research are not that necessary. (ER)

4 What Helped and What Hindered in Achieving Effective Outcomes?

Despite the variation between host institution and context, several common themes emerged from conversations[1] about factors that supported/constrained embedding and collaboration in the ER projects.

---

[1] Embedded researchers in both cohorts had on-boarding conversations with UKCR Champion Kate Lonsdale (the lead author of this paper) and periodic cohort meetings to share progress and discuss what was going well and less well, including the extent to which they were able to embed in their host organisations.

4.1 Being 'on the Inside' of the Organisation

Understanding how their project connected to the wider work of their team (or organisation) helped ERs to see how their project would add value. It also allowed host colleagues to see how they could support the research. The time needed for ERs to understand a different organisational culture, motivation and ways of working was sometimes underestimated, however.

What Helped

Sorting logistics (IT, access to data platforms, HR responsibilities) before the ER was in post; providing an organisational email so they were seen as part of the organisation (both internally and externally); and scheduling time with colleagues from the outset to understand the 'big picture' context of the work (with space for informal conversations).

What Hindered

Delays in accessing IT systems and data platforms; ERs feeling isolated and unsure how to contribute; and key host staff moving on, thereby losing the context and champion for the project.

4.2 Flexibility in the Research Workplan

The process benefitted from being seen as (at least partially) flexible and part of an ongoing enquiry into what had relevance in the problem context. This means not fixating too soon on specific outputs, milestones and timelines, but 'holding them lightly', to be explored as part of a process of mutual learning. Attitudes varied as to how fixed the initial work plan was across the UKCR ER projects; some were concerned about the barrier to future funding if named outputs were not achieved, while others focused on the ultimate goal, assuming that the workplan to achieve it was flexible and an essential part of any collaboration. One ER suggested that collaboratively defining the key questions has value in itself, and particularly valuable if it lays the groundwork for long-term collaboration.

What Helped

Conversation throughout, from initial scoping to final evaluation; and periodically reflecting on actual progress (compared with what had been expected) and the reason for any discrepancies.

What Hindered

Concerns about the workplan being 'fixed' before the work commenced, with inadequate understanding of the host context.

4.3 Openness to Learning on Both Sides

Both hosts and ERs need to be open to continued learning, as supported by the ongoing comparison between 'expectation' and 'what emerges' referred to above. So much of our practical knowledge is tacit, or even unacknowledged and out of our awareness. What seems obvious to the ER may be news to the organisation and vice versa. One ER, for example, reportedly underestimated the extent to which their host understood climate risk, causing a significant revision of the outputs. By allowing time to digest and talk through a problem, outlooks can be gently challenged, connections nurtured and 'aha' moments cultivated—especially when aspects that are only obvious to one side are revealed.

What Helped

Seeing discussions about changes to the workplan or an output as an important aspect of ensuring that the work remains fit-for-purpose; and recognition that there is not always best practice to follow, but that emerging and promising practice is understood and strengthened through dialogue and integrating different types of knowledge.

What Hindered

Pressure for the researcher to have answers before they have had time to understand the context; and pressure to take action before the relevant knowledge is integrated.

4.4 Seniority and Length of Service Are Less Important Than Personality and Outlook

The question "Are early career researchers (ECRs) better suited to this or can researchers at any career stage embed?" was discussed throughout the scheme. Some hosts particularly enjoyed working with someone at an early stage in their career, but others wondered, given the fixed funding available, would fewer days of a more senior academic be of greater value? One ER weighed up the options an ECR could fully embed within an organisation and create high-quality outputs. Conversely, they could be so concerned with developing their academic career that this shapes the focus of the work and they prioritise writing articles over spending the time to truly work for their host. Equally, a more senior person may be more 'set in their ways', used to advising and less good at listening, which might result in poorer outputs. They could have the experience and job security to focus on delivering high-quality outputs with the host.

Reflecting on the range of embedded researchers and the outcomes achieved, the relationship between the level of seniority and the project outcome proved inconclusive. Overall, the UKCR ER experience suggests that personality, skills and outlook are a better guide to impact than career stage.

What Helped

An ER taking a facilitative, person- and situation-centred approach; and an ER that is skilled in listening, making connections, seeing opportunities to achieve common goals and checking for relevance.

What Hindered

An ER imposing their own research agenda; an ER overly concerned with academic career progression; a lack of curiosity about the wider system from the ER's perspective; and an ER with poor communication, teamwork and convening skills.

4.5 Adequate Commitment from ER and Host

The extent to which hosts actively engaged with the process varied considerably, which was in part linked to the host's investment in the research question. Cohort 1 ERs (particularly ECRs with less established networks) valued the two-stage application process (where they could respond to a list of host-identified research needs) and some ERs made clear that they would welcome more mechanisms for practitioners to advertise research needs. This was largely lost for Cohort 2, with the onus on researchers to forge their own connections.

Collaborative hosts found that the ER approach was more time-consuming than traditional research but resulted in more useful outputs. Not all hosts appreciated that for the embedding approach to work well, they needed to be actively engaged throughout the research process—from bid writing to evaluation. The UKCR approach could learn from other embedding approaches, such as FRACTAL [https://www.fractal.org.za/] to develop memorandums of understanding (MOUs) at the start, to clarify expectations, roles and responsibilities of all sides of an ER project.

Few ERs were able to commit full-time to the host setting, largely because of existing teaching commitments. Clearly, ringfencing days for embedding helped, although working from home during lockdown made this differentiation harder.

What Helped

Clear host understanding of, and commitment, to the scheme; ongoing conversations about expectations; and ERs ringfencing sufficient time without other distractions.

What Hindered

Blurring of time-boundaries between competing demands; burnout; working from home; and additional family commitments.

5 Conclusions

The boundary-spanning challenge that the ER projects were intended to accomplish is not one of a simple transfer of knowledge from academic producers to decision-making users, but about building connections between different sources of knowledge, the people (and organisations) who produce and hold this knowledge, and their decision-making processes. To work well, this requires research questions of importance to both host and researcher; it also requires upfront and ongoing dialogue about goals, drivers and outputs, alongside a mutual openness to change plans to accommodate emerging insights and identify more fit-for-purpose outputs that align the different professional, academic and personal contexts of those involved.

The approach worked when it focussed on 'what matters' for the organisation, at a scale that made sense. Adaptation is context specific—organisations need support to explore tentative areas of interest, to understand the implications of headline climate messages for core business, to identify and trial new approaches, and to develop action plans and strategies. Across the 13 projects, the ER scheme provided tailored, human–human support, helping hosts to explore how to respond to climate risk in ways that felt meaningful. In the future, the scheme could be extended to enable organisational staff to embed in academic institutions, to better understand academic knowledge production.

The embedding approach is not a panacea for all circumstances and not the only way to span policy-practice-science boundaries. To work well, it requires a commitment of time, goodwill, flexibility and an openness to learning and 'not knowing' on both sides that can seem counter-cultural in some organisations—including academia where there is pressure to be 'an expert'.

Being an ER means not having 'the answer' but working with others to pool knowledge and experience to produce something that is fit-for-purpose. ERs contextualise their academic knowledge through listening and understanding organisational constraints and incentives. This requires working with others to balance the big picture and the detail, to critically reflect on what has value, to unlearn previous assumptions and to be willing to change course to achieve the most appropriate outcome as new insight emerges. The skills needed to do this well are not yet commonly taught or valued in academia and deserve to be better appreciated and incentivised if we are to address the ongoing disconnect between climate information and adaptation action and ultimately achieve societal climate resilience.

References

1. Oliver T.H., Benini L., Borja, A., Dupont, C., Doherty, B., Grodzi´nska-Jurczak, M., Iglesias, A., Jordan, A., Kass, G., Lung, T., Maguire, C., McGonigle, D., Mickwitz, P., Spangenberg, J.H. and Tarrason, L. 2021 Knowledge architecture for the wise governance of sustainability transitions. *Environmental Science and Polic*, **126**, pp. 152–163.
2. Coles T.C., Hall M. and Duval D.T. 2006. Tourism and Post-Disciplinary Enquiry, *Current Issues in Tourism* **9**(4–5), pp. 293–319.
3. Guston D.H. 2001 Boundary organizations in environmental policy and science: an introduction. *Science, Technology and Human Values* **26**(4), pp. 399–408.
4. Lorenzoni I., Jones M. and Turnpenny J. 2007 Climate change, human genetics, and post-normality in the UK. *Futures* **39**(1), pp. 65–82.
5. Cvitanovic, C., McDonald, J. and Hobday, A.J., 2016. From science to action: principles for undertaking environmental research that enables knowledge exchange and evidence-based decision-making. *Journal of Environmental Management* **183**(3), pp. 864–874.
6. Bednarek, A.T., Wyborn, C., Cvitanovic, C., Meyer, R., Colvin, R.M., Addison, P.F.E., Close, S.L., Curran, K., Farooque, M., Goldman, E., Hart, D., Mannix, H., McGreavy, B., Parris, A., Posner, S., Robinson, C., Ryan, M. and Leith, P. 2018. Boundary spanning at the science–policy interface: the practitioners' perspectives. *Sustainability Science* **13**, pp. 1175–1183.
7. Scott, D. and Taylor, A. 2019. Receptivity and judgement: expanding ways of knowing the climate to strengthen the resilience of cities, FRACTAL Working Paper #7.
8. O'Hare, P. 2021. Manchester's climate risk: a framework for understanding hazards & vulnerability. Manchester, Manchester Climate Change Agency [Online] Available at: https://www.manchesterclimate.com/sites/default/files/Climate%20vulnerability%20framework.pdf.

Open Access This chapter is licensed under the terms of the Creative Commons Attribution 4.0 International License (http://creativecommons.org/licenses/by/4.0/), which permits use, sharing, adaptation, distribution and reproduction in any medium or format, as long as you give appropriate credit to the original author(s) and the source, provide a link to the Creative Commons license and indicate if changes were made.

The images or other third party material in this chapter are included in the chapter's Creative Commons license, unless indicated otherwise in a credit line to the material. If material is not included in the chapter's Creative Commons license and your intended use is not permitted by statutory regulation or exceeds the permitted use, you will need to obtain permission directly from the copyright holder.

PART II

Managing Climate Risks

CHAPTER 5

Putting Climate Resilience in Its Place: Developing Spatially Literate Climate Adaptation Initiatives

Freya Garry, Paul O'Hare, Claire Scannell, Jenna Ashton, Michael Davies, Katy A. Freeborough, Alan Kennedy-Asser, Neil Macdonald, Stephen Scott-Bottoms and Liz Sharp

Abstract

- Understanding the socioeconomic, cultural, historical and political nuances of a place is essential for realising effective local decision-making for climate action.

Lead Authors: Freya K. Garry, Paul O'Hare & Claire Scannell

Contributing Authors: Jenna C. Ashton, Michael Davies, Katy A. Freeborough, Alan T. Kennedy-Asser, Neil Macdonald, Stephen Scott-Bottoms & Liz Sharp

F. Garry (✉)
Met Office, Exeter, UK
e-mail: freya.garry@metoffice.gov.uk

© The Author(s) 2024
S. Dessai et al. (eds.), *Quantifying Climate Risk and Building Resilience in the UK*,
https://doi.org/10.1007/978-3-031-39729-5_5

- People are central to understanding place-based risk and resilience, with consideration of inequality and vulnerability required for effective place-based climate adaptation.
- Temporality is important. Place is not fixed, but changes over time, together with the community that inhabits it.
- Discussing and sharing community knowledge increases the likelihood of successful creation and implementation of climate adaptation practices.
- A sense of place can be deployed to build connections between people, across policy and between scales.

Keywords Adaptation · Place · Resilience · Local · City · Regional

P. O'Hare (✉)
Manchester Metropolitan University, Manchester, UK
e-mail: paul.a.ohare@mmu.ac.uk

C. Scannell (✉)
Met Éireann, Dublin, Ireland

J. Ashton · S. Scott-Bottoms
University of Manchester, Manchester, UK

M. Davies
University College London, London, UK

K. A. Freeborough
British Geological Survey, Nottingham, UK

A. Kennedy-Asser
University of Bristol, Bristol, UK

N. Macdonald
University of Liverpool, Liverpool, UK

L. Sharp
University of Sheffield, Sheffield, UK

1 Introduction

Climate change has profound implications for societies across the world. The impacts of climate change are most acutely experienced at local scales, in the buildings, streets, neighbourhoods, towns and cities where people live and work. It is at this most granular spatial scale that climate change becomes a lived reality.

Unlike climate mitigation, often framed as a global collective effort, adaptive pathways and interventions are dynamic social processes [1], realised at local scales, that require tailoring to specific contexts of people(s) and place(s) in all their complex and intricate assemblages. But implementation challenges abound [2]; successful adaptation requires careful coordination across a myriad of local actors and organisations, often with divergent interests and agendas. Moreover, investments for adaptation can have multiple dividends [3], addressing other pressing local issues—for example by improving our environment (e.g. parks, treescapes and waterways), or providing efficient, reliable, green transportation.

The concept of 'place' is subject to debate and interpretation across disciplines [4]. A key point is how it differs from words like location or resolution, which are common concepts in climate research. Place combines a physical understanding of a location with the social, cultural, sensual and psychological values that people hold. For example, a detached scientific interpretation may suggest that a village has to be abandoned due to its location. However, for the people who live, or have lived or worked in that place, complete abandonment may seem inconceivable as this is where their children grew up, where their ancestors are buried and it holds their memories. The sense of distress and grief associated with environmental change, and with the prospect of relocation is termed 'solastalgia' [5]. Moreover, 'place' is never settled. It is, rather, in a perennial process of becoming [6], continually emergent, negotiated, contested and renegotiated.

This chapter coalesces learning from projects that considered how climate adaptation and resilience have resonated with 'place' across its multifaceted geographies and socioeconomic, cultural and political characteristics. It outlines practical ways of accommodating climate resilience within a place, from working with communities to understand how they use their buildings and what place means to them, to what information helps decision-makers at municipal and neighbourhood scales take climate action.

The chapter emerged from an initial discussion of the importance of place-based climate resilience projects, followed by a workshop session. Case studies from across the UK Climate Resilience Programme (UKCR) are briefly described, revealing how the sense of place changes at different scales but can also serve as a vital connection between us.

Given the complexity and subjectivity of interpretations of the concept, contributors have been given latitude to identify how place, a sense of place, and place attachment emerged as critical themes through their work.

2 Local Places

2.1 ClimaCare

The project 'ClimaCare' quantifies climate related heat risks in care settings nationwide and enhances our understanding of human behaviour, building performance, organisational capacity and governance to enable the UK's care provision to develop equitable adaptation pathways to rising heat stress. For the first time in the UK, ClimaCare collected temperature and humidity data in around 40 care settings, to assess the recurring risk of summertime overheating.

Results from London care homes suggest that overheating is prevalent and prolonged, especially at night; the inability to cool down overnight can lead to increased mortality. Since the severity of overheating was lower in older buildings, building construction age can be a key factor for overheating risk [7].

In analysing differences between bedroom spaces and lounge areas, ClimaCare developed understanding at the sub-building scale, plus insights into how building design and behaviour have changed over time (i.e. a focus on winter warmth in recent years, with little ventilation). The work of ClimaCare, therefore, highlights the usefulness in understanding how buildings—vital constituent elements of place—are constructed, operated and managed in future climate adaptation planning.

2.2 CLandage

The project 'CLandage' uses a historical lens to learn about resilience. By learning how rural communities have adapted and responded to challenges (including extreme weather) in the past, communities are reminded that places evolve over time [8].

One case study was a Grade II listed stone bridge at Pooley Bridge (Cumbria). Built in 1764, the bridge was lost during severe floods in December 2015 and replaced by the UK's first stainless steel bridge. The name 'Pooley Bridge' reflects a longer history; 'Bridge' was added around 1800. Prior to this, the site was known as 'Pooley' or 'Pool How' (the 'pool beside the hill') and was situated at the foot of an ancient Iron Age hill fort. The site is a strategically important bridging/fording place within the wider landscape, with each new bridge designed to accommodate greater flows and protect the community. The bridge's purpose and design have changed through time as the needs of the people in that place changed. The newest steel bridge was designed to consider the wider policy and regional needs of the new UNESCO and Site of Special Scientific Interest (SSSI) designations, as well as respecting local history and environment, and providing functional infrastructure within a working landscape. As well as exhibitions and workshops, learning from this project has been encapsulated in a toolkit for identifying, assessing and characterising river heritage in collaboration with local communities, which will help support decision-making.

3 NEIGHBOURHOODS

3.1 MAGIC

The 'MAGIC' project (https://www.communityactionforwater.org) explored how to sensitively engage local people in helping to hold rainwater back in their neighbourhoods, to both reduce flood risk at times of heavy rain and to enhance understanding of more natural and sustainable drainage. It revealed that engaging the public could augment the extent to which excess water can be stored in the landscape through domestic 'rainwater management systems' (such as water butts or rain gardens) [9]. Additionally, these systems provide an understandable small-scale 'model' of sustainable drainage systems, which were simultaneously being developed in the locality.

As climate change magnifies the risks of pollution and flooding, the need for public understanding of sustainable drainage becomes more urgent in order to enhance community buy-in when authorities propose sustainable drainage on public land, augment the public system by promoting small-scale rainwater management on private properties and increase acceptance for public funding of sustainable drainage.

Community buildings (including a church, a primary school and a general store) were taken as a focus for engagement with local people using arts-based methods, such as drawing and ideas boxes, to stimulate discussions about where rain could be directed and held back. The key question addressed was: "Where around this building can we contain the rain, while also making it a better place for you to be?". A key concern was to discuss rain without giving too much emphasis to flooding, which had locally traumatic associations. MAGIC's community partners set up the cooperative 'Susdrainable' https://www.susdrainable.coop, to both construct the MAGIC rainwater management systems and support the placement of similar systems on other domestic and commercial buildings. The legacy of the project includes the five rainwater management features in the landscape, associated signposting, and the ongoing work of Susdrainable.

3.2 Creative Climate Resilience

The 'Creative Climate Resilience' project focussed on community knowledge and creativity in an ex-industrial electoral ward in Manchester with high levels of social housing. Data on the local population show high incidence of poor health, poverty and social deprivation, low levels of voting, and conflicting development agendas.

Creative Climate Resilience explored how socially engaged arts and community-based performance methods can be used to identify barriers and solutions, articulate perspectives, and offer processes, tools and skills to initiate climate mitigation and adaptation strategies. By taking a political 'ward' boundary, the project offered insights for how local authority practices compare with the city scale. Practically, the project informed the local climate action plan through active collaboration with the local authority and other stakeholders. Innovative outputs included toolboxes, school packs, animations and performance.

Investigating perceptions, knowledge and experiences of 'local' place and neighbourhoods (i.e. distinctiveness, identities, care, activity, networks and physical assets), provided insights around the complexity of community resilience. Artistically, exploration of folklore and storytelling transform the ways in which place and landscapes are perceived and imagined, folding nature and culture together, to capture the ways in which a place is shaped by interwoven social, psychological and topographical factors. The Creative Climate Resilience project has, importantly, participated extensively in local activities in order to understand existing models of collaboration, creative solutions and care, and to identify barriers and opportunities around resilience.

4 Cities

4.1 London Climate Action

The 'London Climate Action' project developed an understanding of how the urban subsurface (the land and built infrastructure below the surface level) could help deliver the City of London's Climate Action Strategy https://www.cityoflondon.gov.uk/services/environmental-health/climate-action/climate-action-strategy. The City of London has a particularly unique sense of place with a small population of residents but a high migrant workforce, coupled with a congested and historic built environment both above and below ground. Access to subsurface data is complex, costly, restricted and sometimes lacking entirely. Subsurface models that work well in some urban spaces are problematic in other locations.

The project recognises policy and data sources relevant to the subsurface in relation to five key climate adaptation measures: (1) sustainable drainage systems; (2) urban greening and tree planting; (3) cool spaces below ground; (4) ground source energy; and (5) prevention of damage to buried utility services. This project found challenges with reconciling data, policy and other issues in different organisations, highlighting a need for government and organisations to work together on data creation, access and model development. The project identified a 'subsurface' dimension to resilience planning, demonstrating gaps/barriers to mapping the use of these places and recommending implementation of suggested adaptation measures.

4.2 Meeting Urban User Needs

The project 'Meeting Urban User Needs' provided a number of UK cities with the capacity to access and interpret local climate information for decision-making.

Cities are complex systems with unique social and economic vulnerabilities, and decision-makers need to navigate climate adaptation in an equitable way. They first need to understand how climate change might impact the city as a whole, as well as the relationship between climate hazards and specific city vulnerabilities, so that at-risk areas and vulnerable groups can be prioritised for adaptation planning.

Urban City Packs were developed with a user journey approach in mind, building awareness, depth of understanding and capacity for implementing informed climate action. Services were developed with cities, and broad climate messaging and climate model output were framed at the local level.

While each city had similar requirements in terms of climate information, every city was unique in terms of capacity to interpret and use information; depending on location, different hazards (e.g. heat, flooding, drought) and vulnerabilities (e.g. infrastructure, health, population) were prioritised, resulting in bespoke products to fit user need. All cities required additional support to understand how they could use the information in their climate action. It was important to work closely with the city, to inform both the design of the city pack (an understanding of 'place') and ongoing development of the service (an understanding of use).

4.3 Manchester Climate Ready

The 'Manchester Climate Action' project [10] demonstrated that at a city scale, climate adaptation policy and practice must be cognisant of—and complement—municipal characteristics of place. This resonance is essential not only to assure acknowledgement by city leaders, but also to gain traction in the congested arenas of public policy and urban governance. Such an approach is vital if climate policy is to be integrated with broader economic, social and cultural policies and practices.

In Manchester—a city with a self-declared vision to be in the top flight of world-class cities by 2025 (manchester.gov.uk/info/500313/the_manchester_strategy/7177/our_vision_for_manchester_in_2025)—it was clear that climate policy should be reconciled with its socioeconomic ambitions. By appealing to these agendas, climate resilience policy not only gains greater recognition in its own right but also has the potential to use broader policies as a vehicle for greater climate resilience.

A placement undertaken with the Manchester Climate Change Agency carefully situated climate policy within these wider contexts, with outputs including the development of a vision (https://www.manchesterclimate.com/content/2022-update) for 'progressive climate resilience' [11] and an associated series of principles that would integrate with other city-wide agendas and ambition. The project has a lasting legacy through on-going work referred to as 'Manchester Climate Ready'.[1]

5 Regions

5.1 Once Upon a Time

The 'Once Upon a Time' project explored how stories and storytelling can effectively be used in climate risk research in Northern Ireland. It considers contrasts in local climate between eastern and western counties, extending the idea of 'place' to the emergence of 'networks between places'.

Initial visual storytelling, in the form of infographics, started with a top-down, data-driven approach, producing risk-related metrics for the agricultural community through an app (https://ukcrp.shinyapps.io/AgricultureNI/). Feedback quickly highlighted the need for further detailed contextual information at a more localised scale, leading to discussion groups being organised across different regions to allow stakeholders to tell their stories. While research often translates hazard data (e.g. number of heatwaves) to impacts (e.g. livestock heat stress) across large spatial scales, local idiosyncrasies mean methods to do this may come with significant uncertainties. Working in collaboration with stakeholders to locally contextualise data can help understand the appropriate 'risk currency' [12] for a given place and help reduce uncertainties that are inevitable in a one-size-fits-all model.

Furthermore, in Northern Ireland, there is a considerable degree of east–west variability in climate impacts, in terms of forage for livestock during hot, dry summers, for example. Province-wide resilience could

---

[1] See Chapter 4 of: https://www.manchesterclimate.com/sites/default/files/2022%20Update%20of%20the%20Manchester%20Climate%20Change%20Framework%20%282020-25%29%20AA.pdf.

potentially be built by strengthening existing networks between places or communities in different geographic locations that may be impacted asynchronously in time.

6 Conclusions

The chapter demonstrates how, across a multitude of distinct case studies and projects, 'place' is key to the extreme weather and climate impacts we experience and can be used to frame climate adaptation and resilience in a more accessible and more efficacious way. We identify below six key principles from these projects to aid climate adaptation by practitioners and policymakers.

1. Climate action and adaptation are more likely to be effective if they acknowledge and are congruent with local senses of place. That said, despite the great importance of place-based actions, these should not be developed at the expense of addressing the wider structural issues that limit both climate action and adaptation, such as the challenge of achieving joined up policy making.
2. Place is relatable. It is where the implications of climate change become personal and tangible. Place, and a sense of place attachment, can be used to evoke the urgency and necessity of adaptation and to energise policymakers at local scales.
3. Place changes over time, as do cultural norms and the behaviours of people that inhabit a place. Set ideas of place can be restrictive for imagining futures or different ways of living. Understanding how a place has evolved and adapted across time and how this will be influenced by climate change is vital, therefore, to understand what will facilitate and inhibit adaptation.
4. Understanding local vulnerability is vital for realising adaptation, such as how buildings, neighbourhoods and wider areas are planned, constructed and function, and how each of those elements relate to each other. Decision-makers must also be attuned to inequality across and between places. Inequalities and inequity within existing decision-making around place is a major barrier for adaptation, and one that must be carefully understood and confronted.
5. Place is not just about the buildings or infrastructure; we must engage with people. Folklore and storytelling can be used to unearth community knowledge and creativity, to identify barriers, solutions

and perspectives, and offer processes, tools and skills to initiate climate action and adaptation strategies. Change happens when communities are united by a common threat or sense of purpose. Developing proposals with local people not only helps shape adaptation to meet local needs, but also enhances local understanding of the need and means to mitigate and adapt to climate change.

6. Place connects not only the people in localities but across different scales. The distinctiveness of particular places and their similarities with other locations provide useful reference points and, potentially, a locus of connectivity. Place is distinct, yet also nested in scale, relational and in connection with other locations.

Acknowledgements Many thanks to our colleagues for their hard work on these UKCR projects, including the rest of the ClimaCare team, Kevin Malone and Sarah Lindley, University of Manchester (Creative Climate Resilience), Christine Sefton and Frances Ellis, University of Sheffield, Dan Galbraith and Kate MacDonald, Timebank Hull and East Riding (MAGIC), Victoria Ramsey and Elizabeth Fuller, Met Office (Meeting Urban User Needs). Thanks, too, to Manchester City Council and the Manchester Climate Agency for contributions to the Manchester Climate Action project. We also thank Candice Howarth and Lucy Vilarkin, and the UKCR Champion Team for their suggestions.

References

1. Adger, W. N. 2003. Social capital, collective action, and adaptation to climate change. *Economic Geography* **79**(4), pp. 387–404.
2. Shamsuddin, S. 2020. Resilience resistance: The challenges and implications of urban resilience implementation. *Cities* **103**.
3. United Nations 2020. UN Common Guidance on Helping Build Resilient Societies. New York, UN.
4. Cresswell, T. 2004. Defining place. Place: A Short Introduction. Malden, MA: Blackwell Ltd, pp. 12.
5. Albrecht, G., Sartore, G.M., Connor, L., Higginbotham, N., Freeman, S., Kelly, B., Stain, H., Tonna, A. and Pollard, G., 2007. Solastalgia: the distress caused by environmental change. *Australasian Psychiatry*, **15**(sup1), pp. S95–S98.
6. Pred, A., 1984. Place as historically contingent process: Structuration and the time-geography of becoming places. *Annals of the Association of American Geographers* **74**(2), pp. 279–297.

7. Gupta, R., Howard, A., Davies, M., Mavrogianni, A., Tsoulou, I., Jain, N., Oikonomou, E. and Wilkinson, P., 2021. Monitoring and modelling the risk of summertime overheating and passive solutions to avoid active cooling in London care homes. *Energy and Buildings* **252**.
8. Naylor, S., Macdonald, N., Bowen, J. P. and Endfield, G. 2022. Extreme weather, school logbooks and social vulnerability: The Outer Hebrides, Scotland, in the late nineteenth and early twentieth centuries. *Journal of Historical Geography* **78**, pp. 84–94.
9. Sefton, C., Sharp, L., Quinn, R., Stovin, V and Pitcher, L. 2022. The feasibility of domestic rain tanks contributing to community-oriented urban flood resilience, *Climate Risk Management* **35**(1–15).
10. O'Hare, P. 2021. Manchester's climate risk: a framework for understanding hazards and vulnerability. Manchester Climate Change Agency. [Online] Available at: https://www.manchesterclimate.com/sites/default/files/Manchester%20Climate%20Risk_A%20Framework%20For%20Understanding%20Hazards%20and%20Vulnerability.pdf.
11. O'Hare, P., 2022. Manchester climate ready: Developing progressive resilience across the city. Manchester Climate Change Agency. [Online] Available at: https://www.manchesterclimate.com/sites/default/files/Progressive%20resilience_FINAL_UPLOAD.pdf.
12. Roberts, F.E., De Meyer, K. and Hubble-Rose, L. 2021. Communicating climate risk: a handbook. Climate Action Unit, University College London. London, United Kingdom.

Open Access This chapter is licensed under the terms of the Creative Commons Attribution 4.0 International License (http://creativecommons.org/licenses/by/4.0/), which permits use, sharing, adaptation, distribution and reproduction in any medium or format, as long as you give appropriate credit to the original author(s) and the source, provide a link to the Creative Commons license and indicate if changes were made.

The images or other third party material in this chapter are included in the chapter's Creative Commons license, unless indicated otherwise in a credit line to the material. If material is not included in the chapter's Creative Commons license and your intended use is not permitted by statutory regulation or exceeds the permitted use, you will need to obtain permission directly from the copyright holder.

CHAPTER 6

Learning from Arts and Humanities Approaches to Building Climate Resilience in the UK

Edward Brookes, Briony McDonagh, Corinna Wagner, Jenna Ashton, Alice Harvey-Fishenden, Alan Kennedy-Asser, Neil Macdonald and Kate Smith

Abstract

- This chapter shares insights from five arts and humanities-led UK Climate Resilience Programme (UKCR) projects, presenting key learnings and pathways for future research and policy interventions.

Lead Authors: Edward Brookes, Briony McDonagh & Corinna Wagner

Contributing Authors: Jenna Ashton, Alice Harvey-Fishenden, Alan Kennedy-Asser, Neil Macdonald & Kate Smith

E. Brookes (✉) · B. McDonagh (✉) · K. Smith
University of Hull, Hull, UK
e-mail: e.brookes@hull.ac.uk

© The Author(s) 2024
S. Dessai et al. (eds.), *Quantifying Climate Risk and Building Resilience in the UK*,
https://doi.org/10.1007/978-3-031-39729-5_6

- We highlight the significant potential of place-based arts and humanities approaches for working with and engaging communities in building climate resilience and driving climate action.
- We underline the importance of generating genuine two-way dialogue, knowledge exchange and co-creation between academics, practitioners, and community members.
- We point to the importance of robustly and reflexively assessing the effectiveness of arts and humanities-led engagement.
- We argue that working collectively to develop more integrated climate and arts/cultural policy is imperative for supporting future long-term climate resilience.

Keywords Arts · Humanities · Community engagement · Climate resilience

1 Introduction

Over the last decade, a growing body of research has identified the effectiveness of arts and humanities approaches for connecting climate science with communities that stand to be most affected by climate change [1–3]. This links to a range of strategies that explore how audiences can be engaged with climate issues through creative, historical and place-based encounters [4–7]. However, the outcomes and opportunities for learning from arts and humanities-based research are not always well disseminated or valued by disciplines outside of arts and humanities circles [7–9]. This includes at national policy level where arts and humanities have exerted little influence over the ways in which climate change is framed

---

C. Wagner (✉)
University of Exeter, Exeter, UK

J. Ashton
University of Manchester, Manchester, UK

A. Harvey-Fishenden · N. Macdonald
University of Liverpool, Liverpool, UK

A. Kennedy-Asser
University of Bristol, Bristol, UK

within public discourse and climate policy—despite targeted recommendations for policymakers [9–11]. As a result, the benefits of mobilising arts and humanities approaches in order to build climate resilience remain underutilised by climate scientists and policymakers [11].

This chapter addresses these research and policy gaps, sharing key learnings from five arts and humanities-led UKCR projects. Here we provide an overview of the evolving academic and practice-based discussions that emerged during the lifespan of each project, and present reflections identified in a series of collaborative workshops with project teams in spring 2022 and subsequently through a paper session at the Royal Geographical Society Annual International Conference in August 2022 [12]. In doing so, this chapter will demonstrate the value of arts and humanities approaches for engaging communities with climate change impacts and action, and the importance of place and dialogue for building effective climate resilience. The projects are as follows:

- 'Community climate resilience through folk pageantry' (Creative Climate Resilience), led by Dr Jenna Ashton at the University of Manchester.
- 'Risky Cities: Living with water in an uncertain future climate' (Risky Cities) and the related UKCR impact project On the Edge, led by Professor Briony McDonagh at the University of Hull.
- 'CLandage: Building climate resilience through community landscapes and cultural heritage' (CLandage), led by Professor Neil Macdonald at the University of Liverpool.
- 'Once upon a time in a heatwave' (Once Upon a Time), an embedded researcher project led by Dr Alan Kennedy-Asser at the University of Bristol.
- 'Time and Tide: Resilience, adaptation, art' (Time and Tide), an embedded researcher project led by Professor Corinna Wagner at the University of Exeter.

What follows is divided into three sections. The first reflects on the importance of place-based approaches in driving awareness, action and resilience building, while the second explores issues around community dialogue. The third focuses on the impacts of these projects, including local and national policy outcomes. The final section summarises our key learnings and suggests directions for future enquiry.

2 Importance of Place

Central to all projects was a belief that place mattered and that place-based approaches help make climate impacts more tangible and relatable to members of the public—and so build a platform for engagement and action [7, 13]. Local stories and place-specific climate messages proved valuable for each of the UKCR projects in being able to transition from small-scale questions about community resilience to larger scale issues such as climate change, so that personal and community resilience was built through understanding past extremes in the local area. Importantly, several projects utilised the local as a lens through which to connect the past, present and future in productive ways to drive anticipatory action. Understanding historical relationships to place and environment proved important for facilitating engagement with generational and longer-term interactions that communities have had with their environment. This helped to generate a sense of identity and environmental continuity that was conducive to positive climate action [14].

The UKCR projects variously harnessed place-based and historically informed approaches, using different geographical lenses and delivering project outcomes at varying geographical scales. 'CLandage' and 'Risky Cities' worked with archival material including maps, civic records, antiquarian histories, diaries and newspapers—for Cumbria, Staffordshire and the Outer Hebrides, and Hull and the East Riding of Yorkshire, respectively—to research experiences of living with climate, weather and flood for use in delivering local climate interventions. By contrast, 'Creative Climate Resilience' worked at a much smaller scale, focusing on the ward of Miles Platting and Newton Heath in Manchester, selected in part because of its high socioeconomic, health and political inequalities and conflicting development agendas. Working within an area defined by its political boundary offered insights into how local authority practices materialise at a micro level in ways that are distinct from city or regional scales. Investigating people's perceptions, knowledge and experiences of 'local' place and neighbourhoods—distinctiveness, care, activity, networks, assets—proved vital for local participation and inclusion in underlining the complexity of community resilience, and what this offers for mitigation and adaptation strategies. In turn, explorations of folklore and mythological storytelling have helped transform the way place and landscape are perceived and imagined in Miles Platting and Newton Heath, folding nature and culture together and promoting a personal connection to climate change that stimulates awareness, action and resilience.

Narratives connected to place are also important to the 'Once Upon a Time' and 'Time and Tide' projects. Memory and anecdote add personal stories to otherwise impersonal data. In 'Once Upon a Time', participants explored their relationship to place and how this intersected with climate-related experiences to generate individual stories. These often tied memories of weather extremes to dates, places, activities or senses, or explored a theme in the past, present and future. The insights were then brought together to produce compelling narratives, as was the case for the 'Future of the Northern Irish Countryside', a story produced in collaboration with local storyteller Liz Weir. The creative act of storytelling provided an alternative way for the climate research community to explore place-based climate data at more intimate scales than is produced by climate risk modelling. The 'Time and Tide' project features large sculptural bells that ring at high tide, installed at sites around the British coast from the Isle of Lewis to Cornwall (Fig. 1). The bells were catalysts for sharing memories about climate change in coastal communities, with each bell a centrepiece for conversations amongst local grassroots groups, educators, regional conservation groups and arts hubs. Participants designed and implemented activities ranging from beach schools to 'TEDx'-style panels with the aim of translating place-based stories into plans of action on climate change. As a result, members of citizen science groups have collaborated with academics and contributed to scientific findings while the Friends of Par Beach and school groups in Harwich, Essex have cleaned beaches.

3 Generating Dialogue

All the projects went beyond addressing specific knowledge deficits or one-way communication, working instead to foster two-way dialogue, knowledge exchange and co-creation between academics, practitioners and community members. They all centre on equity and social justice concerns, working to ensure that communities have agency over the knowledge that they are part of producing and that it is used in ways that are beneficial to them. This was especially important in working with communities whose past experiences may have been of research being 'done on' rather than 'with' them.

Fig. 1 Appledore Time and Tide Bell. Artist: Marcus Vergette (*Photograph* Corinna Wagner, 2021)

Arts and humanities-based approaches offer unique opportunities to facilitate dialogue creation. Creative workshops offer an approachable way for communities to engage in academic research. At the same time, creative practices, especially handwork (such as sewing, knitting and crafting), can offer space for difficult conversations about sea level rise, coastal erosion and loss, particularly when they draw upon place-based and historically informed stories, which make big stories of global change more relevant and legible at the local scale. For 'Risky Cities' and 'CLandage', intensive programmes of archival recovery [15, 16] fed into creative workshops, offering opportunities for participants to work with archival materials, maps and material objects while sharing their experiences of weather, climate and flood. 'Risky Cities' recovered an 800-year history of living with water and flood in Hull and the surrounding region, using these resources to inform a series of place-based, historically informed arts events (see the discussion of FloodLights below) and a community engagement programme involving textile and creative writing workshops, a soundscape and a touring exhibition. 'CLandage' developed workshops and exhibitions that used cultural heritage materials from Staffordshire and the Outer Hebrides to generate dialogue around climate and extreme weather; for example, the project utilising qualitative records of past weather as a prompt for participants to write about their own memories, or to reinterpret the original source material coloured by their own experiences and understanding of the local environment.

In both projects, participants' experiences fed into the research process, culminating in co-created outputs including poetry, [17] creative writing, craft and storytelling that were used in digital and in-person exhibitions curated by and displayed within the communities concerned. Each exhibition also facilitated discussion around climate change between workshop participants and their family and friends. Similarly, with the 'Time and Tide' initiative, local oral histories collected at bell sites (e.g. Morecambe in Lancashire, Redcar in North Yorkshire and Harwich in Essex) were the starting point for multi-artist exhibitions and creative writing publications. Oral histories revealed much about the decline of coastal industries, land erosion and flooding, but also provided insights into the language, images and cultural references that people use to express their feelings and plans. Age-old narratives of 'The Flood[1]' formed the basis of performances by Cornwall-based theatre group Prodigal UPG (https://prodigalupg.com), while video responses to the question "What does the sea mean to you?" were reproduced in the film COTIDAL (https://timeandtidebell.org/cotidal-new/), by artist Tania Kovats.

'Once Upon a Time' also used a simple question to generate dialogue: "What is your favourite thing about the countryside?". Participants were then able to use this prompt to explore their own experiences of place and climate change. 'Creative Climate Resilience' utilised a related but distinct model of socially engaged practice through arts-based research, generating stories, images, performances, music and creative objects (Fig. 2) in order to encourage residents, local authority members, environmental charities, religious organisations, community developers, youth workers and schools to participate in creatively articulating perspectives and solutions for climate mitigation and adaptation issues—and thus feed into local climate action plans. As one contributor to this chapter eloquently

---

[1] The use of 'The Flood' refers to stories of flooding or deluge often attributed to deity or deities, sent to destroy civilisations or punish the wicked, often in an act of divine retribution.

put it, "Researching through embedded engagement brings people with you on a journey of curiosity and knowledge creation, and ensures that both the academic research and creative outputs are genuine public scholarship". Crucially, practice such as this demonstrates how creative research is joyful and playful, while also having serious implications for decision-making.

Similar experiences were reported in On the Edge, a collaborative project between the National Youth Theatre and the University of Hull, funded by a UKCR impact award. The 90-minute, co-created theatrical performance platformed young people's experiences of living with climate change in coastal and estuarine communities on a global policy stage at the 26th UN Climate Change Conference of the Parties (COP26) in Glasgow UK. It comprised a new play by Adeola Yemitan called I Don't Care, and a climate cabaret directed by Tatty Hennessey including spoken word, poetry, music and magic. This rich and stimulating project

Fig. 2 Creating objects for the 'Creative Climate Resilience' project (*Photograph* Jenna Ashton, 2021)

was characterised by intensive two-way dialogue between researchers and young creatives, facilitated via online development workshops and in-person rehearsals. Reflective journals kept by participants—including the academic researchers—chart the cognitive, bodily and emotional experiences of those involved in producing a piece of theatre that critiqued the barriers to climate action experienced by young people, and challenged collective expectations about young people's experiences of the climate crisis.

4 Understanding Community and Policy Impacts

Demonstrating the value of arts and humanities approaches to do more than simply 'window dress' climate science required each of the teams to robustly and reflexively assess the effectiveness of arts and humanities-led engagement to drive climate resilience. This demand was addressed in three ways. First, each of the projects worked to ensure measures for assessing effectiveness were developed in relation to the needs and existing resources within communities, rather than imposed upon them, even though the precise measures of success used varied across the projects. In the case of 'Creative Climate Resilience', effectiveness was understood as a scale where outcomes were identified by and with the participants. For some participants, that meant empowering them to join the conversation around climate action; for others it meant exploring how they could move from climate or political apathy to awareness and action. Effectiveness also included being able to provide new insights for decision-makers and developing new processes, content and storytelling that contributed to existing resilience activities and supported the community to thrive not just survive. These outcomes were then measured utilising an embedded process that identified the individual and organisational changes amongst those involved in the project, as well as the legacy projects that emerged from engaging in the research process.

At the same time, the project teams recognised that persuading policy audiences of the value of arts-based engagement in driving climate action—and thereby increasing the uptake of these approaches—is facilitated by being able to chart (and on occasion, quantify) our impacts. 'The Risky Cities' team, for example, analysed audience feedback from its programme of community arts interventions including 'FloodLights' (Fig. 3), a series of multimedia, light and sound installations exploring Hull's experiences of living with water past, present and future, which took place in Hull city centre in October 2021 and attracted an audience of more than 11,000. Survey responses demonstrate that the event drove shifts in people's thinking about living with water, flooding and climate change, with 64% of respondents reporting that the event made them think about climate futures, and a third reporting behavioural changes they planned to make in relation to this.[2] As the survey results make clear, place-based approaches—in particular, site-specific installations that mobilised Hull's watery histories and identities—were crucial in generating engagement and action towards climate resilience.

Finally, the UKCR projects discussed here each made direct policy interventions. 'Creative Climate Resilience' centred their approach around the co-design of an open access toolkit for different interest groups to be able to undertake their own climate action planning. This was embedded across a wider spectrum of knowledge exchange with local authority actors—including neighbourhood teams, climate officers, elected members, community groups and scrutiny committees—that all fed into the local climate action plan. This was also a reflexive process, documenting policy engagement as the project progressed and sharing processes and findings. Similarly, 'Risky Cities' targeted a range of local and national policy audiences to shape best practice for resilience building through arts and humanities. This included hosting a climate resilience workshop for local stakeholders; developing a policy brief shared with MPs across Hull and the local council; contributing to flood risk policy (e.g. Hull City Council's Local Flood Risk Management Strategy for 2022–2028, POSTNOTE 647 on Coastal Management [11, 18, 19]; and tabled amendments to the Levelling Up Bill by Hull MP, Emma Hardy); and contributing to cultural policy (e.g. responses to the Department for Digital, Culture, Media and Sport inquiry on culture, place-making and the levelling up agenda) [20].

5   Conclusions

In conclusion, we highlight three key learnings and two directions for future research and policy interventions and, in doing so, advocate for a specifically arts and humanities approach to climate resilience that centres

---

[2] Based on 457 survey responses.

Fig. 3 Audience members enjoying the Sinuous City installation, part of the FloodLights event in Hull (*Photograph* Briony McDonagh, 2021)

on people and their experiences and helps us to rethink what resilience means at the local, community scale.

First, the projects collectively underline the significant potential of place-based, arts and humanities approaches—including those drawing on learning histories—to raise awareness, drive climate action and build climate resilience. These approaches make complex scientific ideas meaningful and big global narratives tangible at the local level, supporting people to understand what complex climate futures might mean for them. Second, we highlight the importance of generating genuine dialogue and co-creation, rather than one-way communication about climate futures. The projects here exemplify varying approaches and possibilities, but all sought to grant community stakeholders and policymakers the agency and urgency through which to act and inform future resilience building strategies. Third, all the projects stress the importance and the difficulties of assessing the 'effectiveness' of arts and humanities-led approaches. They push us to think about what successful engagement means, while also recognising that measurable outcomes—whether expressed qualitatively or quantitatively—are important in persuading others about the value of arts and humanities-led approaches for climate resilience.

Relatedly, our research and policy engagements have also identified important knowledge gaps which must be addressed if the full impacts of arts and humanities-led climate interventions are to be realised. As all the UKCR arts and humanities-led projects show, future long-term resilience plans need to respond effectively to the local cultural and place-specific impacts of climate change. Working collectively to develop more integrated climate and arts policy is, therefore, imperative in supporting this. At the same time, current national cultural policy prioritises the economic value of arts and heritage events [21]. Future policy needs to go beyond this and recognise both intangible benefits of arts engagement *and* its importance for addressing climate concerns and building resilience [22]. We look forward to working collectively with policymakers, climate scientists, community stakeholders and other actors in embracing these challenges.

Acknowledgements We would like to thank Professor Stephen Scott-Bottoms and Dr Hannah Fluck for providing valuable insight and support in the formulation of this chapter.

Each of the projects would like to thank their respective research teams, project partners and community participants for their involvement and valuable contributions throughout. We would also like to acknowledge the financial contributions of the UK Climate Resilience Programme, UK Research and Innovation (UKRI) and the Arts and Humanities Research Council (AHRC) in supporting each of the projects involved.

REFERENCES

1. Moser, S. 2019. Not for the faint of heart: Task of climate change communication in the context of societal transformation. In Feola, G., Geoghegan, H. and Arnall, A. (Eds.). *Climate and Culture: Multidisciplinary Perspectives of Knowing, Being and Doing in a Climate Change World*. Cambridge University Press, Cambridge, pp. 141–168.
2. Moser, S. 2016. Reflections on climate change communication research and practice in the second decade of the 21st century: what more is there to say? *WIREs Climate Change* 7, pp. 345–369.
3. Hulme, M. 2011. Meet the humanities. *Nature Climate Change* 1, pp. 177–179.
4. Scott-Bottoms, S. and Roe, M. 2020 Who is a hydrocitizen? The use of dialogic arts methods as a research tool with water professionals in West Yorkshire, UK. *Local Environment*, pp. 1–17.
5. Rice, R., Rebich-Hespanha, S. and Zhu, H. 2019. Communicating about Climate Change Through Art and Science. In Pinto, J., Gutsche, R. and Prado, P. (Eds.), *Climate Change, Media and Culture: Critical Issues in Global Environmental Communication*. Emerald Publishing Limited, Bingley, pp. 129–154.
6. Burke, M., Ockwell, D. and Whitmarsh, L. 2018. Participatory arts and affective engagement with climate change: The missing link in achieving climate compatible behaviour change?. *Global Environmental Change* 49, pp. 95–105.
7. Corbett, J. and Clark, B. 2017. The arts and humanities in climate change engagement. In von Storch, H. (Ed.). *Oxford Research Encyclopaedia of Climate Science*. Oxford University Press, New York, pp. 1–17.
8. Honeybun-Arnolda, E. and Obermesiter, N. 2019. A Climate for Change: Millennials, Science and the Humanities. *Environmental Communication* 13(1), pp. 1–8.
9. Blue, G. 2016 Framing Climate Change for Public Deliberation: What Role for Interpretative Social Sciences and Humanities?. *Journal of Environmental Policy and Planning* 18(1) pp. 67–84.

10. Clark, K. 2021. Policy Review: Valuing Culture and Heritage Capital: A Framework Towards Informing Decision Making. *The Historic Environment: Policy and Practice,* **12**(2), pp. 252–258.
11. Julies Bicycle. 2021. Culture: The Missing Link to Climate Action. [Online] Available at: https://juliesbicycle.com/policyportal/internationalpolicy.
12. Session title: Understanding Climate Resilience through the Arts and Humanities. Contributors included: Neil Macdonald, Alice Harvey-Fishenden, Briony McDonagh, Kate Smith, Edward Brookes, Jenna Ashton and Simon Naylor
13. Marschütz, B., Bremer, S., Runhaar, H., Hegger, D., Mees, H., Vervoort, J. and Wardekker, A. (2020). Local narratives of change as an entry point for building urban climate resilience. *Climate Risk Management* **28**, pp. 1–15.
14. Lee, K. 2021 Urban public space as a didactic platform: Raising awareness of climate change through experience arts. *Sustainability* **13**(5), pp. 2915.
15. Naylor, S., Macdonald, N., Bowen, J. and Endfield G. 2022. Extreme weather, school logbooks and social vulnerability: The Outer Hebrides, Scotland, in the late nineteenth and early twentieth centuries. *Journal of Historical Geography* **78**, pp. 84–94.
16. McDonagh, B., Mottram, S., Worthen, H., and Buxton-Hill, S. (in review) 'Living with water and flood in medieval and early modern Hull'.
17. Wardle Woodend, M., Harvey-Fishenden, A. and Macdonald, N. 2022. Flood and Drought Poetry: Experiences of Weather Extremes in Staffordshire. Staffordshire: Dreamwell Writing Limited. [Online] Available at: http://www.dreamwellwriting.simplesite.com/.
18. Hull City Council. 2022. Local Flood Risk Management Strategy 2022–2028, *Hull City Council,* UK. [Online] Available at: https://www.hull.gov.uk/sites/hull/files/media/Editor%20-%20Planning/Appendix_1_-_Legislative_and_policy_context.pdf.
19. UK Parliament POST. 2021. *Coastal Management.* [Online] Available at: https://post.parliament.uk/research-briefings/post-pn-0647/.
20. Smith, K., McDonagh, B., Brookes, E. and Pilmer, L. 2022. Reimagining where we live: cultural placemaking and the levelling up agenda. Department for Digital, Culture, Media & Sport (DCMS) Call for Evidence.
21. Lawton, R., Fujiwara, D., Arber, M., Maguire., H, Malde, J., O'Donovan, P., Lyons, A. and Atkinson, G. 2020 Department for Digital, Culture, Media & Sport (DCMS) Rapid Evidence Assessment: Culture and Heritage Valuation Studies - Technical Report, *Department for Digital, Culture, Media and Sport,* UK. [Online] Available at: https://assets.publishing.service.gov.uk/government/uploads/system/uploads/attachment_data/file/955142/REA_culture_heritage_value_Simetrica.pdf.

22. Clark, K. 2021. Policy Review: Valuing Culture and Heritage Capital: A Framework Towards Informing Decision Making. *The Historic Environment: Policy and Practice,* **12**(2), pp. 252–258.

Open Access This chapter is licensed under the terms of the Creative Commons Attribution 4.0 International License (http://creativecommons.org/licenses/by/4.0/), which permits use, sharing, adaptation, distribution and reproduction in any medium or format, as long as you give appropriate credit to the original author(s) and the source, provide a link to the Creative Commons license and indicate if changes were made.

The images or other third party material in this chapter are included in the chapter's Creative Commons license, unless indicated otherwise in a credit line to the material. If material is not included in the chapter's Creative Commons license and your intended use is not permitted by statutory regulation or exceeds the permitted use, you will need to obtain permission directly from the copyright holder.

PART III

Tools for Resilience Building

CHAPTER 7

What Have We Learned from the Climate Service Projects Delivered Through the UK Climate Resilience Programme?

Caitlin Douglas and Mark Harrison

Abstract

- Climate service delivery depends on the presence of positive enabling conditions within service providers, user organisations and the wider context in which the prototype is being developed (i.e. the political, economic, social, cultural or legal landscape).
- User trust in a service output is critical; direct engagement through co-production can help build this trust, facilitated by managing

Lead Authors: Caitlin Douglas & Mark Harrison

C. Douglas (✉)
King's College London, London, UK

M. Harrison (✉)
Met Office, Exeter, UK
e-mail: Mark.Harrison@MetOffice.gov.uk

© The Author(s) 2024
S. Dessai et al. (eds.), *Quantifying Climate Risk and Building Resilience in the UK*,
https://doi.org/10.1007/978-3-031-39729-5_7

expectations and clearly communicating service scope and limitations.
- The ambition to scale up climate services remains challenging, in part due to limitations within existing funding frameworks (particularly in relation to building relationships with new sectors), plus a lack of ongoing support for users.

Keywords Climate services · Resilience · Adaptation · Co-development

1 Introduction

Climate services aid the effective use of climate information by individuals, businesses, government and other organisations. They are increasingly used due to growing public and political awareness of the need to take action to adapt to climate change [1].

Global, regional and national efforts are under way to facilitate the development and uptake of climate services, such as the Global Framework for Climate Services (https://gfcs.wmo.int/), the European Roadmap on Climate Services [2] and the UK Climate Resilience Programme (UKCR).

This paper provides an overview of the projects UKCR delivered, touches on the challenges they faced and the successes experienced, and highlights lessons that have been learned along the way.

2 Overview of Projects

Fifteen climate services projects were commissioned through UKCR, with half focussed on prototyping and the other half on improving other aspects of climate service delivery (e.g. developing standards and approaches to valuing services).

Prototypes were trialled in a range of contexts, exploring new research and new markets, as well as working with established users to address known challenges. A large proportion of the projects commissioned focused on providing climate information to the water sector (mature and heavily regulated), as well as the urban sector (emerging).

Table 1 provides an overview of the projects. Project outcomes (i.e. how the services are used) are more difficult to articulate due to the time lag between service delivery and impact, the inability of the user[1] to articulate the impact, or an unwillingness/inability of the user to share this information. However, the expectation is that the outputs will be helpful to users in some way, such as informing policy development, changing management practices, raising awareness, or providing new guidance, workflows or evidence base.

3   Key Learnings

To gather insights into the projects outlined in Table 1, we conducted one-hour, semi-structured interviews with representatives from each of the project teams (25 April–31 May 2022). We took notes, identifying key topics arising from the interviews. These topics were then compared and themes identified through discussion. Due to time constraints, we were not able to engage beyond service providers. While most project leads had a good understanding of their users, it would have been insightful to obtain a user voice directly. This might have highlighted issues with usability that were not fully appreciated by the service provider.

Common themes emerged from our research, which we have categorised as follows: enabling environment, user trust, and scalability. The key learnings relating to these three topics are discussed below.

3.1   *Enabling Environment*

This theme refers to the environment in which a prototype is being developed — does it help or hinder the prototyping process? During our discussions, it became apparent that this is a critical aspect, which consists of (1) the provider organisation, (2) the user organisation and (3) the wider context (political, economic, social, cultural or legal) in which the prototype is being developed. This relationship is shown in Fig. 1.

---

[1] Although we use a singular term 'user' we recognise that users are not a homogenous unit but rather represent a variety of types of organisations (from individuals to large companies), and that there exists considerable variation in knowledge and information needs within organisations themselves. User refers to both end users and intermediaries.

Table 1 Climate services projects commissioned by the UK Climate Resilience Programme. We define 'applied research' as capability-led development with subsequent user engagement, 'prototype development' as exploratory work to develop new products for new markets, 'product development' as the creation of new products in response to clear user needs, and 'delivery' as being additional activities to support the development and delivery of climate services. We define sector maturity as: low (no regulation), medium (regulation but minimal engagement), or high (active engagement with regulation)

Project title (abbreviated)	Project type	User sector	Sector maturity scale	Aims	Intended user/s
Health Sector Resilience	Applied research	Health	High	Provision of datasets and tools to enable the public health sector to estimate future health-climate risks	NHS, UK Health Security Agency
FUTURE-DRAINAGE	Applied Research	Water	High	Provision of datasets to enable users to design suitable drainage to withstand high rainfall	Environment Agency, Scottish Environment Protection Agency, water companies, flood risk management authorities, academia, infrastructure owners
Coastal Climate Services	Applied Research	Coastal	Low	Provision of datasets, tools and information to support stakeholder decisions around coastal resilience	Environment Agency, professional bodies
OpenCLIM	Applied Research	N/A	N/A	Provision of an open framework for assessing climate change risks and adaptation options in a consistent manner across sectors, including the tools to enable the coupling of models	Department for Environment, Food & Rural Affairs, local government, service providers

Project title (abbreviated)	Project type	User sector	Sector maturity scale	Aims	Intended user/s
Meeting Urban User Needs	Prototype development	Urban	Low	Provision of information to enable urban stakeholders to showcase what a changing climate means for their location	Local government
Transport/Energy Climate Services	Prototype development	Transport & Energy	Medium	Provision of information to enable transport sector managers to assess the resilience of their assets and operations	Department for Transport, arm's length bodies
Freshwater Monitoring & Forecasting	Product development	Water	High	Provision of a service that provides the evidence required to allow stakeholders to make both short- and longer-term decisions	Environment Agency, water companies, Scottish Water, NatureScot
Water Sector Resilience	Product development	Water	High	Provision of data and tools to enable water companies to meet their regulatory obligations and better plan for future droughts	Water companies
FREEDOM-BCCR	Product development	Water	High	Provision of information to support investment decisions	Water companies, Water Services Regulation Authority (Ofwat)

(continued)

Table 1 (continued)

Project title (abbreviated)	Project type	User sector	Sector maturity scale	Aims	Intended user/s
Bristol Heat Resilience*	Delivery	Urban	Low	Provision of information to local council to support the development of strategies to build resilience and reduce heat risk	Local government
Creative Climate Resilience	Delivery	Urban	Low	Development of a process to help build climate resilience within marginalised communities	Local government, residents, community groups
Climate Stress Testing	Delivery	Food	Low	Knowledge brokered between the earth observation sector and those in the UK food supply chain to improve space-enabled climate services	Private sector, national government
Upscaling Climate Service Pilots	Delivery	N/A	N/A	Provision of a toolset to help climate service providers to scale their impact more effectively	Service providers
Climate Resilience Standards	Delivery	N/A	N/A	Provision of information to assess how climate services standards could be developed for UK authorities, companies and academia	Service providers

Project title (abbreviated)	Project type	User sector	Sector maturity scale	Aims	Intended user/s
Climate Services Standards	Delivery	N/A	N/A	Co-development of a climate services standard that helps both climate services providers and users demonstrate that the climate service is trustworthy and accessible	Service providers, end users, standards bodies, World Meteorological Organization, World Bank
UK-SSPs	Delivery	N/A	N/A	Production of a set of shared socioeconomic pathways (SSPs) for the UK climate resilience community to inform further research about the UK's risk and resilience to climate change	Researchers

(continued)

Table 1 (continued)

Project title (abbreviated)	Project type	User sector	Sector maturity scale	Aims	Intended user/s
National Framework for Climate Services	Delivery	N/A	N/A	Creation of a shared vision for a UK National Framework for Climate Service together with a set of recommendations on its structure and implementation	National government
Climate Information for Decision-making	Delivery	N/A	N/A	Provision of a set of recommendations and associated evidence base that will be used to decide on key priorities for future climate modelling and science in the UK	Service providers, end users

*We were unable to speak to a representative from this project due to a change of personnel

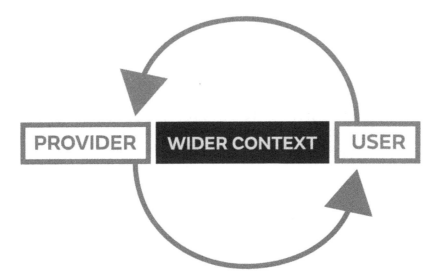

Fig. 1 The enabling environment in which a prototype is being developed affects its development and delivery. Conditions within the provider organisation, the user organisation and the wider context in which they operate all need to incentivise climate service delivery

3.1.1 Provider Organisations

The following key learnings emerged on how the focus, expertise and organisational structure of service providers affect climate service development:

- Definition of success: Typical metrics of 'success' within many organisations (e.g. publications, sales, downloads) are not appropriate for incentivising climate services development. This is because they bear no relation to whether a service is used or even beneficial to the intended user. This disconnect made it difficult to understand the impact of a service on users and to monitor and evaluate success. Consideration is needed to define and measure success within service providers.
- Skills: Many climate service providers have a background in research; therefore, staff are experienced in delivering peer-reviewed journal

articles, datasets and reports. However, staff may be less experienced in user engagement and service development; they, therefore, may have more limited knowledge regarding establishing user needs, communicating science and supplier capabilities, and delivering actionable services. As usability, usefulness and utility often drive the uptake and value of services, service providers would benefit from recognising and valuing the skills needed to facilitate these attributes. Similarly, the skillset required to co-develop with users tends to be under-appreciated in projects, yet critical for success. See Chapter 3 for further details on the need for (and challenges of) co-developing climate services.

3.1.2   User Organisations

The following key learnings emerged on how aspects within the user organisation can influence service development. It should be noted that these were drawn from conversations with the project leads (not end users).

- Relationship: The relationship between the service provider and the user organisation is an important element in service development [3]. Those we spoke with indicated that co-development tended to be more effective when a principal contact stayed in post and engaged throughout. Progress increased when the principal contact had the remit, time and willingness to engage with both the service provider and internally within the user organisation. Funding and project planning tends to focus on the climate service provider rather than the user; the time and effort required from the user organisation is often overlooked, or resources are assumed (e.g. time, energy, knowledge, funding).
- Sector: The nature of the sector is important. A highly competitive and more regulated market appears to result in users being more receptive to engaging with climate services. This was evident in the water sector, where users were willing to engage with UKCR projects—either for competitive advantage or for fear of being left behind. In a heavily regulated sector such as this, there is a clearer pathway to align climate service provision with existing regulation (see 'Wider context' section below).

3.1.3 Wider Context

The following key learnings emerged on how the wider political, economic, social, cultural or legal context affects the ability of service providers and user organisations to engage with climate services.

1. Funding landscape: Many of the funded projects were well defined from the start as they built on previous working relationships and projects. This is likely a direct result of the nature of the funding calls: limited funding and project duration meant successful proposals tended to have a clear scope and corresponding benefit. These external pressures disincentivise expansion into new users or sectors, as it takes time to develop relationships and understand end-user requirements. We recognise that funding bodies are required to demonstrate impact and therefore need to be confident that a project will deliver; but this pressure to achieve tangible impact within a limited timeframe appears to deter the development of services for new sectors.

    Similarly, many of the barriers to service provision partly stem from the approach used to commission the work, typically focussing on asking the service provider to consider what capability could be developed to meet a customer need rather than taking a genuinely user-centric approach. Also, the difficulty in so many projects articulating outcomes may be a function of the commissioning process as documenting outcomes was not built into project design. Many side benefits (e.g. network/relationship development) were cited as outcomes but not recorded as part of the project documentation.

2. Regulation: Regulation can drive action (see section on 'User organisations' above). This is because users are clear of the minimum requirement placed upon them, and service providers have a clear steer as to what information is required. By extension, regulatory requirements to undertake climate action could facilitate climate service uptake, whereby suppliers understand information need and users are provided with the remit to engage with suppliers. For example, reporting required as mandated by the Task Force on Climate-related Financial Disclosures (https://www.fsb-tcfd.org/) (TCFD) and the Climate Change Act 2008 provides a helpful enabling environment.

3.2 User Trust

Projects employed a variety of approaches to build relationships with users. User trust in a service output is critical. Building relationships with users and involving them in service development are well-established approaches, as is clear communication about service scope and limitations, being careful not to over-sell current or future service provision.

Many of the projects were built on pre-existing relationships, which meant that the co-development process could start swiftly. Interactions between service providers and users varied considerably, in terms of the platform (e.g. emails, WhatsApp, workshops, events, one-to-one meetings) and frequency (e.g. ad-hoc, weekly, quarterly). While the COVID-19 pandemic was cited as having a negative impact on engagement (e.g. staff shortages), there were also several positive reflections, such as greater familiarity with remote working and collaboration.

The timing of user engagement also varied. Some advocated for engagement from project inception, while others saw value in developing a demonstrator before initiating engagement. In the case of the latter, the provider organisations had a good appreciation of user need, so were less likely to take a fundamentally wrong development path. However, while there are benefits to focussing discussions on a tangible product, it runs the risk of making incorrect assumptions about users' needs or constraining their thinking.

3.3 Scalability

Developing new services is often best done in a focussed 'bespoke' way through engaging with a limited set of users. This is often followed by an aspiration to reach a wider audience by rolling out similar services to a wider set of users. This concept of delivering climate services at scale was raised in many of the interviews and represents a challenge within the climate service community [4].

While the 'create once, use many' approach might be cost-effective from a delivery perspective, evidence pointed to the following potential drawbacks:

1. Evidence suggests that climate services created without significant user input are used less and potentially misused; therefore, 'scaling' may represent a false economy.

2. To enhance the usability of the climate services, it was reported that language used within the products should align with language used by users and outputs should align with existing metrics.
3. In many cases, it was also reported that generic outputs are ineffective; they fail to provide enough information for those seeking more technical detail, while also failing to be accessible for those seeking a more top-level summary.

The interview insights raise questions about how to feasibly scale climate services, given the need for bespoke outputs. We reflect on the implications of this and other insights in the next section.

4 Implications for Future Climate Services Landscape

The three key themes—enabling environment, user trust, scalability—emerging from our interviews are connected, as there is a clear link between the importance of the enabling environment in increasing user engagement and trust. In turn, both themes have implications for scalability, as bespoke services are more likely to be trusted and used. Based on our findings, we propose a series of recommendations on how to incentivise and scale up climate services.

4.1 Incentivisation

The following recommendations are characterised by stakeholder type.

4.1.1 Service Providers
Service providers should produce outputs that are useful and usable to end users. The following internal mechanisms can be used to facilitate this aim: (1) focus the underpinning scientific work on problems that users need to have addressed; and (2) recognise and reward the work required to make useful and usable climate services, as career paths for these individuals are not as defined as for climate scientists (i.e., publishing articles) or operational meteorologists (i.e., accreditation) (https://www.rmets.org/professional-development).

For (1), lessons could be learnd from the private sector where the funding model is directly related to user satisfaction. For (2), greater external professional recognition of the skills required to co-develop, deliver and support climate services (e.g. through chartership) would be beneficial.

4.1.2 Users

Users should implement ways to work across institutional silos and to embed climate adaptation and resilience into decision-making. There is a role for 'knowledge translators' in bridging silos within and between organisations, which is increasingly recognised in both the public and private sectors.

4.1.3 Context (Regulators)

Regulation, whether from government, industry bodies or international organisations, promotes user engagement with climate services. A key question is how to encourage the benefits of regulation into other sectors. Potential mechanisms include moving towards net zero, stress testing and resilience building, climate exposure reporting (i.e. TCFD) and expansion of the Climate Change Act 2008's Adaptation Reporting Power (https://assets.publishing.service.gov.uk/government/uploads/system/uploads/attachment_data/file/182636/report-faq-110126.pdf) to other organisations such as local authorities.

4.2 Context (Funders)

Through certain projects, UKCR intended to facilitate the co-development of climate services, building on climate research. Recommendations for future programmes of this nature would be to consider following a value stream approach that starts with the user. This would ensure a better linkage between the user and capability development, resulting in a clearer understanding of the overall impact of the programme. We see two priority areas for primary research: (1) developing scientific approaches that provide robust evidence at the level of decision support required by users (see Chapter 8 for further discussion); and (2) focussing on systems, so that interdependencies between sectors and/or risks are captured.

4.3 Scaling up

In this final section, we focus our recommendations on service providers and those within the wider context (note: users are not a key barrier in this regard).

4.3.1 Service Providers

Several projects published data, reports and code in the public domain. While this may seem a transparent and traditional way to share knowledge, there are potential problems, as existing users may not have the technical skills to use the information and new users may not know where to find it. A support system should be put in place to ensure that outputs are not misused, and feedback loops established to shape future development by providers. For services to endure in a meaningful way, it is necessary to consider the business model once the project completes and funding ends—as mandated by the UKSA International Partnership Programme (https://www.spacefordevelopment.org/ipp/), for example.

4.3.2 Context (Funders)

There is a clear need to reach users across a wide range of sectors. However, it takes time to develop new relationships, to understand their needs and to develop the science and services to meet those needs. Current funding approaches tend to require certainty and demonstrable value for money at the proposal stage, whereas it would be helpful to have more of a staged approach to encourage working with new sectors. Directing funding towards the following activities would particularly help to progress service development: (1) improve our understanding of what happens with existing services, to learn how stakeholders use the services and how their needs evolve; (2) consider how existing work could be repurposed or extended to new use cases; and (3) take the next steps towards 'operationalising' services at scale.

5 Conclusions

UKCR made significant progress in furthering the use of climate services in the UK, by working with new user groups, creating new methodologies to enhance climate service delivery, piloting products and developing guidance and resources to support climate service delivery. Barriers remain, however, notably surrounding the development and delivery of climate services. The effectiveness of this process depends on the enabling conditions—the actions of regulators, funders and government (the wider context) ultimately influence the actions of service providers and users. To progress climate services, attention should, therefore, focus on fostering the necessary enabling conditions in these groups.

Acknowledgements We thank those with whom we initially spoke about their respective projects, namely Jenna Ashton, Dan Bernie, Steven Chan, Andrew Charlton-Perez, Daniel Cotterill, Chris Counsell, Murray Dale, Jennifer Dicks, Galia Guentchev, Nicola Golding, Peter Hunter, Don Monteith, Erika Palin, Rachel Perks, Craig Robson, Christophe Sarran, Claire Scannell, Jon Stenning and Louise Wilson.

References

1. Intergovernmental Panel on Climate Change (IPCC). 2022. Climate Change 2022: Impacts, Adaptation and Vulnerability. Contribution of Working Group II to the Sixth Assessment Report of the Intergovernmental Panel on Climate Change [H.-O. Pörtner, D.C. Roberts, M. Tignor, E.S. Poloczanska, K. Mintenbeck, A. Alegría, M. Craig, S. Langsdorf, S. Löschke, V. Möller, A. Okem, B. Rama (eds.)]. Cambridge University Press. Cambridge, UK and New York, NY, USA, pp. 3056.
2. European Commission. 2015. *A European research and innovation roadmap for climate services*. Directorate-General for Research and Innovation. Publications Office.
3. Vincent, K., Daly, M., Scannell, C. and Leathes, B. 2018. What can climate services learn from theory and practice of co-production? *Climate Services* **12**, pp 48–58.
4. Lu, J., Lemos, M.C., Koudinya, V. and Prokopy, L.S. 2022. Scaling up co-produced climate-driven decision support tools for agriculture. *Nature Sustainability* **5**, pp. 254–262.

Open Access This chapter is licensed under the terms of the Creative Commons Attribution 4.0 International License (http://creativecommons.org/licenses/by/4.0/), which permits use, sharing, adaptation, distribution and reproduction in any medium or format, as long as you give appropriate credit to the original author(s) and the source, provide a link to the Creative Commons license and indicate if changes were made.

The images or other third party material in this chapter are included in the chapter's Creative Commons license, unless indicated otherwise in a credit line to the material. If material is not included in the chapter's Creative Commons license and your intended use is not permitted by statutory regulation or exceeds the permitted use, you will need to obtain permission directly from the copyright holder.

CHAPTER 8

What Insights Can the Programme Share on Developing Decision Support Tools?

Rachel Perks, Craig Robson, Nigel Arnell, James Cooper, Laura Dawkins, Elizabeth Fuller, Alan Kennedy-Asser, Robert Nicholls and Victoria Ramsey

Abstract

- The definition of decision support tools in the context of climate change and adaptation is explored, highlighting the variation in approaches to design and form of tools.
- Several challenges are identified that have impeded the successful development of decision support tools, including financial restrictions, time constraints and meaningful stakeholder engagement.

Lead Authors: Rachel Perks & Craig Robson

Contributing Authors: Nigel Arnell, James Cooper, Laura Dawkins, Elizabeth Fuller, Alan Kennedy-Asser, Robert Nicholls & Victoria Ramsey

R. Perks (✉) · L. Dawkins · E. Fuller · V. Ramsey
Met Office, Exeter, UK
e-mail: rachel.perks@metoffice.gov.uk

© The Author(s) 2024
S. Dessai et al. (eds.), *Quantifying Climate Risk and Building Resilience in the UK*,
https://doi.org/10.1007/978-3-031-39729-5_8

- We highlight a number of potential areas for future research, including work to address the challenges of scaling up decision support tools and stronger frameworks for guiding stakeholder engagement.

Keywords Decision support tools · Climate hazard · Adaptation · Stakeholder engagement

1 Introduction

To minimise the risk from the impacts of climate change, both mitigation and adaptation strategies will be required, hence decision-makers—such as government departments, local councils and private businesses—are increasingly interested in potential options to reduce their exposure to climate-related risks. Key enabling tools here are decision support tools (DSTs).

The UK Climate Resilience Programme (UKCR) funded several projects that focused on developing DSTs, where a DST allows users to derive critical information-such as climate hazard to subsequent risk—to make informed decisions. This could be a synthesis of large datasets (making data more digestible), or an interactive tool that displays

---

C. Robson (✉)
Newcastle University, Newcastle upon Tyne, UK

N. Arnell
University of Reading, Reading, UK

J. Cooper
University of Liverpool, Liverpool, UK

A. Kennedy-Asser
University of Bristol, Bristol, UK

R. Nicholls
University of East Anglia, Norwich, UK

climate hazard information alongside associated impacts and mitigation/adaptation options. DSTs take a multitude of forms, from integrated assessment frameworks to visualisation platforms. This paper will focus on the science-user interface and how information to inform decisions is presented.

In broad terms, a DST can be defined as a tool or knowledge resource to support the decision making process, by facilitating a comparison of different climate futures or adaptation options [1, 2] or by enabling information awareness at spatial scales [3]. The way in which this is interpreted allows for the generation of non-uniform, heterogeneous tools, which have been designed for specific use cases and stakeholders. Consequently, the definition of DSTs varies among the climate resilience community, with different expectations among user communities.

In what follows, we discuss some of the key findings with respect to the development of DSTs from across the UKCR programme, including challenges and gaps in understanding to inform future work. This chapter complements other chapters in this collection, including but (not limited to) chapters 11, 7 and 3.

2 Survey and Review of Decision Support Tools

A series of surveys and reviews of projects funded through the UKCR programme were conducted, focusing on projects where an output was regarded as a decision support tool. In total, nine structured interviews were conducted (each 30–60 minutes in length) with project leads. A summary of the DSTs and key stakeholders is provided in Table 1.

2.1 Web-Based Interactive Tools

At a local scale, the UKCR project 'Catchment Erosion Resilience' designed a pilot web-based interactive DST to illustrate changes in erosion risk within rivers under UKCP18 projected extreme rainfall events. Focusing on a single river, the pilot demonstrated the change in erosion risk to critical infrastructure, including roads, bridges, water and waste treatment structures and electricity transmission towers as well as agricultural land. Similarly, to assess future heat risk across a city, part two of the 'Heat Service' (Meeting Urban User Needs) project combined heat hazard information with socioeconomic data to develop a heat vulnerability index (HVI) for Belfast and Hull. This was delivered to users through a web-based ArcGIS StoryMap, allowing users to interact with

Table 1 A summary of the projects interviewed for the survey of UKCR decision support tools, including a description of each tool, the spatial scale it operates on and the stakeholders directly involved in its development

Title of project	Decision support tool	Platform	Scale	Key stakeholders
Climate Risk Indicators	Interactive website allowing visualisation of climate risk indicators across the UK at varying spatial scales and allowing download of data. Primary aim is to assist in raising awareness of potential climate changes	Interactive web tool (https://uk-cri.org/)	National, with varying reporting spatial scales	Wide range of users, including Environment Agency and The Wildlife Trusts
Catchment Erosion Resilience	Interactive visualisation of flood and erosion risks, and associated economic damage to infrastructure, for different rainfall events under UKCP18 climate change scenarios	Interactive web tool (pilot) (https://arcoes-dst.liverpool.ac.uk/EHRC/bootleaf-master3/index_DST_F2.php?map_no=9)	Local	Water companies, electricity transmission and erosion control industry
Risk Assessment Frameworks	Interactive web tool to develop the capability to use climate data in open-source risk assessment framework software, to quantify future climate risk in the UK, explore adaptation option appraisals and assess sensitivities [4]	Interactive web tool—R Shiny (proposed)	National	Department for Education, Ministry of Justice
Once Upon a Time	Interactive web tool exploring changing temperatures for different climate scenarios	Interactive web tool (Northern Ireland Rural Heat Map) (https://akaresearch.shinyapps.io/ruralheat/)	Regional	Climate Northern Ireland, Dale Farm dairy cooperative and Ulster Farmers' Union

(continued)

Table 1 (continued)

Title of project	Decision support tool	Platform	Scale	Key stakeholders
CoastalRes	Prototype methods to assess coastal resilience to erosion and flooding under climate change scenarios at local to national (England) scales [5]	Result datasets, reports, interactive educational web tool (https://coastalresilience.uk/crm/)	Varying spatial scales	Environment Agency and maritime local authorities who manage coastal flood and erosion hazards in England, plus other stakeholders interested in shoreline management planning (e.g. Natural England)
OpenCLIM	An integrated cross-sectoral assessment tool for climate impacts and adaptation options, including hazards such as heat, flooding and water supply, and impacts on people, property, agriculture and biodiversity to support national climate risk assessment as exemplified by the UK's third Climate Change Risk Assessment (CCRA3). The tool is critically underpinned by an open modelling framework which allows for production of new results and updating of workflows and models	Result datasets, modelling framework, interactive web tool (proposed)	National	Department for Environment, Food and Rural Affairs (Defra), Climate Change Committee (CCC) Environment Agency, Climate Ready Clyde, Natural England, Norfolk Broads National Park Authority and many more

(continued)

Table 1 (continued)

Title of project	Decision support tool	Platform	Scale	Key stakeholders
Coastal Climate Services	Part one—A globally relocatable tool to provide regional sea-level projections rooted in the Coupled Model Intercomparison Project Phase 5 (CMIP5) model simulations and Monte Carlo approach, for the future emissions scenarios used in the Intergovernmental Panel on Climate Change's 5th Assessment Report (IPCC AR5). These are based on Representative Concentration Pathways (RCPs) [6, 7] Part two—A dataset of projected future still water Return Levels (RLs) at 2km spacing around the UK coastline, for the future emissions scenarios used in the IPCC AR5 and based on RCPs	Part one—Python-based tool[1] (will be made accessible on completion of the UKCR programme) Part two—Result dataset in GIS format (https://ukclimateprojectionsui.metoffice.gov.uk/products/form/MS4_ESL_Subset_01 and https://ukclimateprojectionsui.metoffice.gov.uk/products/form/MS4_ESL_Subset_02)	Part one: Local Part two: National	Environment Agency, Scottish Environment Protection Agency, National Resources Wales, Department for Infrastructure Rivers (Northern Ireland), flood risk practitioners and Institution of Mechanical Engineers

(continued)

---

[1] Stored as a GitHub repository of Python code.

Table 1 (continued)

Title of project	Decision support tool	Platform	Scale	Key stakeholders
Meeting Urban User Needs (City Packs)	Fact sheets and infographics that use probabilistic projections from UK Climate Projections alongside other information to help inform decision-makers about their climate risks	Infographics, PDF fact sheets	National	Local and city councils
Meeting Urban User Needs (Heat Service)	Part one—A set of factsheets building understanding of heat hazards and impacts in cities Part two—Heat vulnerability index combining climate, socioeconomic and built environment data to assess future heat risk across the city	Infographics, PDF fact sheets, GIS layers, GIS StoryMap	National	Local and city councils, emergency planning groups
Bristol Heat Resilience[2]	Heat Vulnerability Index to explore where heatwaves could have the biggest impact and a Heat Resilience Plan to support the development of green infrastructure strategies	Interactive web tool (https://bcc.maps.arcgis.com/apps/instant/portfolio/index.html?appid=986e3531099f48d393052fab91ceff51)	Local	Bristol City Council

the HVI maps. The tool allows for a narrative to be built around the data and generates maps to aid understanding.

Web tools at larger spatial scales have also been developed; the project 'Once Upon a Time' has enabled users in Northern Ireland to examine likely changes in temperature—and therefore temperature extremes—as shown in Fig. 1. This tool was designed in conjunction with Climate

---

[2] The Bristol Heat Resilience project has been included in the table for information but was not part of the nine interviews undertaken for this paper.

Northern Ireland and others (e.g. agricultural associations) so that more informed decisions can be made.

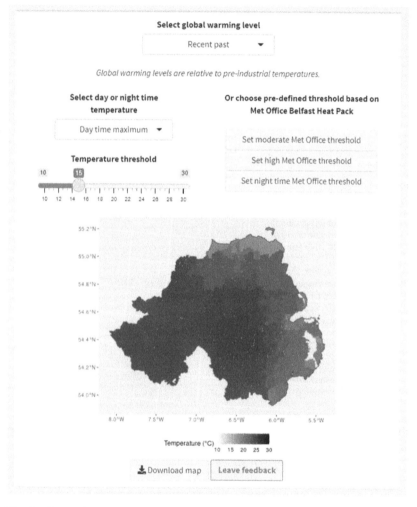

Fig. 1 From the UKCR project 'Once Upon a Time', an example of an interactive DST, which allows users to explore the changes in temperature rise across Northern Ireland over time (*Source* https://akaresearch.shinyapps.io/ruralheat/)

Finally, 'Climate Risk Indicators' developed a web-based interactive DST to provide information on climate risk indicators across the UK at spatial scales ranging from local to national. The indicators cover a range of sectors and are calculated from the latest UK Climate Projections (UKCP18).

2.2 Infographics and Climate Hazard Information

The 'City Packs' (Meeting Urban User Needs) project used the UKCP probabilistic projections to create static fact sheets, using infographics, to raise awareness of the headline messages on climate hazards (such as temperature, rainfall and sea-level rise) likely to affect the given city or region. They were co-developed with relevant authorities to explain the science and the results in a simple, easy to understand format that could be easily distributed.

Based on this success, part one of the 'Heat Service' (Meeting Urban User Needs) project also developed a set of factsheets focusing on heat hazards and associated impacts in cities, to support local and city councils in their decision making around climate change adaptation and to inform planning for future heat events (see example in Fig. 2). Although not interactive, they are highly visual and meet users' needs by providing information on climate specific themes in an accessible and policy-relevant manner.

2.3 Data Outputs

'OpenCLIM' developed a large set of data outputs, covering a range of climate hazards and associated impacts, such as heatwaves, drought and flooding, providing a large resource for information-driven decision making. Additionally, the modelling framework is open and usable for stakeholders, given appropriate training. The output data will allow users to explore these varied hazards and the effects of different adaptation scenarios.

The dataset compiled in part two of the project 'Coastal Climate Services' allows users to explore and download larger datasets of extreme water levels to derive their own understandings of risk, as well as the potential implications of different policy or planning decisions (where applicable data is available). Similarly, the underlying data and spatial outputs in 'Climate Risk Indicators', and the shapefiles produced for the

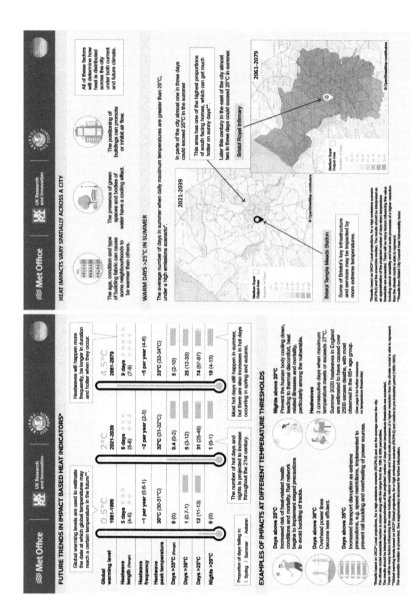

Fig. 2 Example pages from the Heat Pack for Bristol, one of the products of the 'Heat Service' (Meeting Urban User Needs) project. These PDFs are designed to inform stakeholders on possible effects of climate change to help inform decision making (*Source* https://www.metoffice.gov.uk/binaries/content/assets/metofficegovuk/pdf/research/spf/ukcr_heat_pack_bristol.pdf)

heat mapping element of 'Heat Service' (Meeting Urban User Needs) are available for users to download and integrate into internal GIS systems or other software. These approaches allow users to explore the datasets in detail but require an in-depth scientific knowledge for successful application.

3   Development of Decision Support Tools

Developing DSTs requires a level of stakeholder engagement or co-production if they are to be successful [8]. Decision support tool development varied across the UKCR projects, with approaches falling into two broad categories: (1) science-led, where the tool was initially developed prior to engaging with stakeholders to create a bespoke version; and (2) user-led, where stakeholders were engaged from the outset and at predefined intervals throughout the development process.

The 'Risk Assessment Frameworks' project initially developed the DST as a natural next step in the climate risk assessment framework, using the quantified risk to compare adaptation options. Discussions with stakeholders subsequently helped shape the framework beyond a simple cost-benefit analysis tool, to incorporate analyses such as how much a given adaptation action meets organisational objectives. Similarly, 'Climate Risk Indicators' produced a web platform primarily to disseminate project results, with its form and functionality informed by discussions with stakeholders. 'Coastal Climate Services' developed its sea-level tool informed by published literature, with a view to improving its capability and usability following a user engagement workshop (October 2022).

In contrast, the 'City Packs' and 'Heat Service' (Meeting Urban User Needs) discussed the overarching aims and the objectives of the project through larger workshops and smaller meetings, plus follow-up questionnaires. Furthermore, the 'OpenCLIM' project ran workshops both on a regional and per-sector basis to establish stakeholder needs across different groups, from administrative area policymakers to sector experts, through online interactive discussion sessions.

Both science- and user-led approaches to DST development can be successful, leading to purposeful engagement with and uptake of tools. The main advantage of science-led DSTs is that the climate change information is scientifically traceable and robust; however, they do not always filter through to local impacts, possible adaptation options or decision making. The perceived level of success of a DST will depend on what the user need is. For example, if a user requires spatial hazard information for their own impact models such as that provided in the 'Coastal Climate Services' project then this approach can be considered successful; however, if they require additional adaptation information then this approach may not meet the user demand.

A user-led approach has the advantage that the resulting DST is understandable and appropriate for the decision in question, thereby increasing the likelihood of stakeholder uptake. However, the level of scientific rigour can sometimes be called into question when aspects are tailored to suit the user requirements rather than the science. Furthermore, there are issues with workshops which tend towards a broad audience, with a wide range of conflicting thoughts, ideas and interests and so diluting potential learning and beneficial outcomes. Much of the research across the UKCR programme highlights the benefits of co-production over consultation in this regard.

4 Usability of Decision Support Tools

DSTs have maximum impact and effectiveness when designed in collaboration with stakeholders [9, 10], specifically when they are demand-driven rather than science-driven [11], though the design of interactions with stakeholders must be considered otherwise DSTs can still prove to be ineffective [3, 12, 13]. Many of the UKCR-derived tools have been developed through engagement with stakeholders to ensure the DST meets their needs and can inform decision making processes. These tools have been developed in collaboration with groups such as interested local authorities ('City Packs'), government departments ('Climate Risk Indicators', 'Coastal Climate Services') and industry associations ('Once Upon a Time').

An important aspect of usability is fitness for purpose of the DST. For example, the 'Coastal Climate Services' project provides regional sea-level projections, which could help inform investment strategies based on the comparative risk to different regions. However, the information would not be sufficient for making infrastructure development plans, because additional site-specific information would be required to complete a detailed risk assessment, using for example a hydrodynamic model.

The success of DSTs in the context of the UKCR programme can be measured via their continued use, plus their influence on the development of new or existing climate change policies. Reporting of this nature is weak or hard to identify, however. Within the time frame of the UKCR programme, success of DSTs is difficult to measure because many of the projects have only recently been completed. A longer-term assessment, which records evidence from stakeholders where tools have directly influenced policy changes, would allow for more useful evaluation. Certainly, there are several examples where this is anticipated, such as the 'Heat Service' (Meeting Urban User Needs) project, which will be used to update severe weather plans for heat risk in Belfast, as well as the city's climate change risk assessment. Further evidence can be gathered from emerging follow-on projects; for example, 'Catchment Erosion Resilience' produced a pilot tool for examining future river erosion impacts, which has led to other projects with infrastructure providers (e.g. Welsh Water and the National Grid).

5 Barriers in Decision Support Tool Development

Several difficulties have arisen in the UKCR programme for those projects developing DSTs, commonly with respect to user engagement. Firstly, knowing who to approach within potential stakeholder organisations or groups to initialise engagement. Secondly, knowing which method of user engagement is appropriate for the desired outcome (large workshops can garner initial interest, while smaller, more targeted meetings are beneficial when refining the final product). Finally, knowing how to maintain purposeful engagement and manage expectations. To some extent, involving the right users begins to address this, although the relationship needs to be carefully managed to avoid stakeholder fatigue. The recent increase in the use of technology for virtual meetings has made engagement easier, but a balance needs to be maintained between user and developer expectations. Clear frameworks around managing expectations and engagement methods need to be made apparent from the outset of the process to ensure all parties benefit from the engagement, and no one party is left disappointed or frustrated.

Another recognised issue for DSTs is securing legacy access for stakeholders through online portals. Given the amount of resource invested in the development of these useful and usable tools, ensuring DSTs remain available beyond the funding window is especially important. Similarly, ongoing web support and updates are often not possible and knowledge loss may occur when the original project team moves on. This remains a widespread issue affecting how tools remain available and supported for users. Overcoming this is a particular challenge but could be limited in some instances through extensive and detailed levels of co-production, or by ensuring some degree of handover of developed tools to others who are interested in their long-term availability—such as partners or larger central bodies who may have more flexible resources and may want to ensure invested resources, knowledge and technical advancement is not lost from the domain.

6 Conclusions

Several key insights have been learned with respect to DSTs across the UKCR projects. Firstly, DSTs vary in form, based on factors such as the 'decision' they were intended to address, the resources available and the amount of stakeholder engagement. They can be information only or a more complex interactive tool. Secondly, user engagement is often a challenge, despite the pivotal role it plays in DST design. Throughout the UKCR programme, user engagement for the development of DSTs has taken various forms, from targeted meetings to large workshops. Ensuring a suitable and diverse mix of stakeholders is crucial (see chapter 3 for further discussion on co-production and user engagement).

Further, DSTs can be science- or research-led, although fundamentally are dependent on the available science and thus data. Conveying the science is ultimately the role of a DST, although time and rigour is required from a science perspective to achieve this, which is not always fully recognised by stakeholders in the early stages of DST development. Lastly, key barriers to developing DSTs remain, namely funding, skills and legacy planning. Developing effective DST tools requires expertise across the science, usability and visualisation domains, yet few projects have the available resources or skills to do so. Ensuring legacy is an enduring issue.

Along with these insights, we suggest a number of areas of further research to help address key areas of understanding which could be improved in the context of DSTs and climate resilience. As discussed throughout, stakeholder engagement remains a challenge and the development of frameworks which can support this critical activity are essential to smooth this process for researchers, developers and stakeholders alike. These frameworks should consider many elements of the development process, but a further recognised gap that may aid this process is the potential for more consistent visualisation methods for climate-based

DSTs and in particular visualisations of uncertainties in data. Forming a common platform, or set of methods, such as for the visualisation of uncertainty common in climate-based modelling, may reduce time required for stakeholders to understand datasets and thus maximise the time that can be focused on other elements of DSTs and engagement. While some work has been done in this area (e.g. UKCR-funded studies [14, 15]), none of the DSTs reviewed here incorporated existing best practice or consulted visualisation experts to help convey the uncertainties.

Upscaling regional- or city-scale DSTs poses several challenges: from the underlying data potentially being limited geographically, whether due to licensing or because of the devolved management of datasets in the UK (e.g. 'OpenCLIM'); to the increasing data volumes becoming an issue for storage and processing methods; to analysis models being specific to an area (e.g. 'City Packs'). Some UKCR-funded work on addressing the challenges of upscaling has been undertaken through the project 'Upscaling Climate Service Pilots' (see chapter 7), though further work to gauge success regarding DSTs is required. Finally, throughout this review, being able to quantify the success of DSTs has been a consistent challenge given the various forms of tools, intended use and the diversity of stakeholders. The development of approaches to better capture the success and failures of current DSTs would enable future projects to learn from this and subsequently implement changes in their user engagement process and development of such tools.

Acknowledgements Many thanks to those who have contributed through making time for conversations on experiences of developing DSTs within the UKCR programme.

References

1. Palutikof J.P., Street R.B. and Gardiner E.P. 2019a. Looking to the future: guidelines for decision support as adaptation practice matures. *Climatic Change* **153**, 643–655.
2. Palutikof J.P., Street R.B. and Gardiner E.P. 2019b. Decision support platforms for climate change adaptation: an overview and introduction. *Climatic Change* **153**, 459–476.
3. Rodela, R., Bregt, A.K., Ligtenberg, A., Pérez-Soba, M. and Verweij, P. 2017. The social side of spatial decision support systems: Investigating

knowledge integration and learning. *Environmental Science and Policy* **76**, pp. 177–184.
4. Dawkins, L. C., Bernie, D. J., Lowe, J. A. and Economou, T. 2023. Assessing climate risk using ensembles: A novel framework for applying and extending open-source climate risk assessment platforms. *Climate Risk Management* **40**, 100510.
5. Townend, I.H., French, J.R., Nicholls, R.J., Brown, S., Carpenter, S., Haigh, I.D., Hill, C.T., Lazarus, E., Penning-Rowsell, E.C., Thompson, C.E.L. and Tompkins, E.L. 2021. Operationalising coastal resilience to flood and erosion hazard: A demonstration for England. *Science of The Total Environment* **783**, 146880.
6. Palmer, M., Howard, T., Tinker, J., Lowe, J., Bricheno, Calvert, D., Edwards, T., Gregory, J., Harris, G., Krijnen, J., Pickering, M., Roberts, C. and Wolf, J. 2018. UKCP18 Marine report. Met Office.
7. Palmer, M. D., Gregory, J. M., Bagge, M., Calvert, D., Hagedoorn, J. M., Howard, T., Klemann, V., Lowe, J. A., Roberts, C. D., Slangen, A. B. A. and Spada, G. 2020. Exploring the drivers of global and local sea-level change over the 21st century and beyond. *Earth's Future* **8**, e2019EF001413.
8. Lu, J., Lemos, M.C., Koundinya, V. and Prokopy, L.S. 2022. Scaling up co-produced climate-driven decision support tools for agriculture. *Nature Sustainability* **5**, pp. 254–262.
9. Clar C. and Steurer R. 2018. Why popular support tools on climate change adaptation have difficulties in reaching local policy makers: Qualitative insights from the UK and Germany. *Environmental Policy and Governance* **28**, pp.172–182.
10. Wong-Parodi, G., Mach, K.J., Jagannathan, K. and Sjostrom, K.D. 2020. Insights for developing effective decision support tools for environmental sustainability. *Current Opinion in Environmental Sustainability* **42**, pp. 52–59.
11. Capela Lourenço, T., Swart, R., Goosen, H. and Street, R. 2016.The rise of demand-driven climate services. *Nature Climate Change* **6**, pp. 13–14.
12. McIntosh B.S., Ascough, J.C., Twery, M., Chew, J., Elmahdi, A., Haase, D., Harou, J.J., Hepting, D., Cuddy, S., Jakeman, A.J., Chen, S., Kassahun, A., Lautenbach, S., Matthews, K., Merritt, W., Quinn, N.W.T., Rodriguez-Roda I., Sieber, S., Stavenga, M., Sulis, A., Ticehurst, J., Volk, M., Wrobel, M., Delden, H., El-Sawah, S., Rizzoli, A. and Voinov, A. 2011. Environmental decision support systems (EDSS) development: challenges and best practices. *Environmental Modelling and Software*, **26**(12), pp.1389–1402.
13. McIntosh, B.S., Seaton, R.A.F. and Jeffrey, P. 2007. Tools to think with? Towards understanding the use of computer-based support tools in policy relevant research. *Environmental Modelling and Software* **22**, pp. 640–648.

14. Kause, A., Bruine de Bruin, W., Fung, F., Taylor, A. and Lowe, J. 2020. Visualizations of Projected Rainfall Change in the United Kingdom: An Interview Study about User Perceptions. *Sustainability* **12**(7), pp. 2955.
15. Kause, A., Bruine de Bruin, W., Domingos, S., Mittal, N., Lowe, J. and Fung, F. 2021. Communications about uncertainty in scientific climate-related findings: a qualitative systematic review. *Environmental Research Letters* **16**(5).

Open Access This chapter is licensed under the terms of the Creative Commons Attribution 4.0 International License (http://creativecommons.org/licenses/by/4.0/), which permits use, sharing, adaptation, distribution and reproduction in any medium or format, as long as you give appropriate credit to the original author(s) and the source, provide a link to the Creative Commons license and indicate if changes were made.

The images or other third party material in this chapter are included in the chapter's Creative Commons license, unless indicated otherwise in a credit line to the material. If material is not included in the chapter's Creative Commons license and your intended use is not permitted by statutory regulation or exceeds the permitted use, you will need to obtain permission directly from the copyright holder.

PART IV

Understanding and Characterising Risk

CHAPTER 9

Improved Understanding and Characterisation of Climate Hazards in the UK

Jennifer Catto, Simon Brown, Clair Barnes, Steven Chan, Daniel Cotterill, Murray Dale, Laura Dawkins, Hayley Fowler, Freya Garry, Will Keat, Elizabeth Kendon, Jason Lowe, Colin Manning, David Pritchard, Peter Robins, David Sexton, Rob Shooter and David Stephenson

Abstract

- This chapter describes new methods and datasets, developed through UK Climate Resilience Programme (UKCR) projects, to better understand climate hazards.

Lead Authors: Jennifer Catto & Simon Brown

Contributing Authors: Clair Barnes, Steven Chan, Daniel Cotterill, Murray Dale, Laura Dawkins, Hayley Fowler, Freya Garry, Will Keat, Elizabeth Kendon, Jason Lowe, Colin Manning, David Pritchard, Peter Robins, David Sexton, Rob Shooter & David Stephenson

---

J. Catto (✉) · D. Stephenson
University of Exeter, Exeter, UK
e-mail: j.catto@exeter.ac.uk

© The Author(s) 2024
S. Dessai et al. (eds.), *Quantifying Climate Risk and Building Resilience in the UK*,
https://doi.org/10.1007/978-3-031-39729-5_9

- We describe projections of hazards using new tools and provide examples of applications for decision-making.
- Going forward, this new physical and statistical understanding should be incorporated into climate risk assessments.

Keywords Climate hazards · Flood · Extreme weather · High-resolution modelling · Statistical methods for extremes

1 Introduction

To improve resilience in response to climate change, it is vital to have the best possible understanding of weather and climate hazards facing the UK both now and in future. The latest IPCC report [1] used the concept of 'climate impact-drivers' (CID), which are particular climate states that may or may not lead to hazards and subsequent impacts, depending on the global location [2]. This is only one part of the consideration of impacts, since the vulnerability and exposure also need to be considered. Here, we specifically consider a subset of CID that are expected to have

---

S. Brown (✉) · D. Cotterill · L. Dawkins · F. Garry · W. Keat · E. Kendon · J. Lowe · D. Sexton · R. Shooter
Met Office, Exeter, UK
e-mail: simon.brown@metoffice.gov.uk

C. Barnes
University College London, London, UK

S. Chan · H. Fowler · C. Manning · D. Pritchard
Newcastle University, Newcastle upon Tyne, UK

M. Dale
JBA Consulting, Exeter, UK

E. Kendon
University of Bristol, Bristol, UK

P. Robins
Bangor University, Bangor, Wales, UK

negative impacts on the UK namely key hazards of extreme precipitation, high winds and heat extremes. Through the work of the UKCR programme, the data and methods used to identify and characterise these hazards—such as physical and statistical models—have advanced significantly. Researchers now have a better physical understanding of relevant hazards (e.g. how extreme rainfall interacts with different weather systems), which will underpin projections of the three key hazards in the UK. These new insights also enable researchers to better distil and communicate climate information (and associated uncertainties) to decision-makers.

In this chapter, we will discuss each of these aspects and provide a list of datasets and tools described (see Table 1). For further detail on how hazard information has been translated into decision-relevant knowledge, please refer to Chaps. 10 and 11.

2 Advances in Hazard Data

Numerical climate models, which simulate future weather based on our knowledge of the physics of the climate system, are used to simulate future hazards in the UK. Two simulations of the UK Climate Projections (UKCP18), which allow a detailed investigation of the UK and Europe, were used in several UKCR projects: UKCP Regional projections (12 km resolution) and UKCP Local projections (2.2 km resolution). Studies found that the local simulations are superior to the regional simulations, when compared with observational datasets for certain variables [12]. For example, the effects of conurbations on extreme temperatures in the present climate are found to be more realistically simulated for UKCP Local compared with UKCP Regional [13], and the representation of seasonal mean as well as short duration heavy rainfall events is also better [12]. Further, estimates of future river flooding using the DECIPHeR hydrological model for two benchmark catchments (Rivers Thet and Dyfi) differ considerably between UKCP Local and UKCP Regional [14], implying that high-resolution space–time varying precipitation fields are important in flood risk analysis.

To complement the UKCP18 data, the project 'ExSamples' [11] has provided a large and rich set of extreme winter scenarios for the late twenty-first century, using an atmosphere-only model at 60km resolution, run on distributed computing via https://www.climateprediction.net for climateprediction.net. The atmosphere is forced with prescribed sea

Table 1 The table provides UKCR outputs relating to hazards, as described in this chapter: tools and code (blue); websites for data exploration (yellow); and datasets (green)

Project	Product	Resources
STORMY-WEATHER	Front identification code	Ref: Sansom and Catto [3] Webpage: https://github.com/phil-sansom/front_id
FUTURE-DRAINAGE	RED-UP Rainfall Perturbation Tool	Ref: Dale [4] https://ukwir.org/How-do-we-achieve-zero-uncontrolled-discharges-from-sewers-by-2050#case-studies
Improving Climate Hazard Information	HOTdays tool	Ref: Brown [5] Contact: simon.brown@metoffice.gov.uk
Stochastic Simulation	Stochastic weather generator	Contact: hayley.fowler@newcastle.ac.uk
SEARCH	Risk of compound flooding map	Ref: Lyddon et al. [6] Webpage: https://www.researchgate.net/publication/363166162_Historic_Spatial_Patterns_of_Storm-Driven_Compound_Events_in_UK_Estuaries
EuroCORDEX-UK	Data explorer webpage	Webpage: https://github-pages.ucl.ac.uk/UKCORDEX-plot-explorer/
Multiple Hazards	Case studies of agricultural compound hazards	Ref: Garry et al. [7]
STORMY-WEATHER	Storm type dataset	Ref: Catto et al. (in prep) [8] Contact: j.catto@exeter.ac.uk
FUTURE-DRAINAGE	Design rainfall flood uplifts	Ref: Chan et al. (under review) [9] Webpage: https://artefacts.ceda.ac.uk/badc_datadocs/future-drainage/FUTURE_DRAINAGE_Guidance_for_applying_rainfall_uplifts.pdf
Improving Climate Hazard Information	Return levels at high resolution	Ref: Shooter and Brown (under review) [10]
ExSamples	Extreme winter scenarios	Ref: Leach et al. [11]

surface temperatures (SSTs) from unusually warm (Fig. 1) or wet future winters, allowing robust sampling of extreme atmospheric states. This information is potentially suited to informing adaptation planning and decision-making—including high risk scenarios that have impacts across multiple sectors and regions of the UK,—such as infrastructure damage. Aspirations for future work in this area include using higher resolution models and potentially converting these samples into a product for use in UKCP.

Hydrodynamic modelling is required to translate multiple meteorological hazards, such as extreme high sea levels and rainfall, into a compound flooding hazard. The 'SEARCH' project has developed 20–50m scale

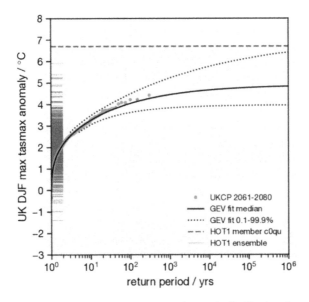

Fig. 1 Return period diagram (From Leach et al. [11]) showing the return period of UK DJF maximum surface temperature anomaly from the UKCP ensemble from 2061–2080. The black curve shows the median of the generalised extreme value (GEV) model fit and the dotted lines show the 0.1–99.9% confidence interval on the GEV fit. The thin orange lines on the left show the UK DJF maximum surface temperature anomalies from the ExSamples ensemble. This figure demonstrates that using the relatively low-resolution model in this study, forced with the SSTs from an extreme hot winter, produces even larger anomalies than the UKCP ensemble. Reproduced according to the CC-BY licence

digital elevation models (that include flood defence heights) for 12 UK estuaries, which are used to parameterise flood inundation models in the landscape evolution model CAESAR-Lisflood. These models have demonstrated that sub-daily river discharge and sea level data are required to understand flood risk in UK estuaries [6, 15, 16] and have established local-scale thresholds for the drivers of flooding. Historical compound flooding across 126 estuaries of Britain was investigated [6] at sub-daily scale to isolate catchments that are currently most vulnerable to storm-driven compound events, such as many of the steep catchments on the west coast of Britain. These tools, together with future projections of extreme rainfall and sea level variability, have been used to determine the future compound flooding hazard for UK catchments and will be available as a risk map.

Physical modelling of climate hazards requires significant computing resource, whereas statistical modelling offers a potentially less expensive alternative, as explored through the UKCR embedded researcher 'Stochastic Simulation' project and the (non-UKCR) 'FUTURE-STORMS' project. This has improved accessibility, openness and usability of a well-established spatio-temporal stochastic weather generator, including extreme value modelling of rainfall improvements. Functionality has also been improved, making it easier for users to run the weather generator with climate perturbations from UKCP18—and helping to overcome barriers to use, thereby allowing a wider range of users to conduct their own simulations for bespoke future hazard characterisation.

A new approach (and dataset) has been developed that allows user-specified heatwave definitions, including day of year and degree of global warming [5]. This allows for the precise characterisation of the severity, duration and frequency of heatwaves for any date and time of year and estimates the probability of heatwaves that are more extreme than any observed or simulated in climate models.

3 Advances in Methods for Characterising Hazards

The accurate characterisation of a hazard is necessary for rigorous risk estimates and appropriate adaptation, but by construction is very difficult due to the rarity of very extreme events. Therefore, robust statistical methods, that make the best use of all available data, are required. A new approach to estimating extreme properties involved pooling data from multiple sites to aid extremal inference at a given location, without losing spatial detail and differences between sites [17, 18]. Using a 'generalised additive model' framework, extreme value distributions are fitted to observations from the UK station network—together with smooth covariates that are functions of elevation, latitude and longitude—to produce extreme value distributions for the whole of the UK at a resolution of 1km. Values for daily precipitation [10] and annual wind gusts were produced at 10, 50 and 100-year return levels. These are available online.

The UKCR project 'Multiple Hazards' has also used generalised additive models to build robust event sets from UKCP18 data, to help quantify probabilities of both uni-variate and multi-variate extreme events occurring by creating larger datasets with the same spatial and temporal statistical properties [19]. The project has also identified how multi-variate climate hazards may impact UK agriculture [7], helping the farming sector understand its adaptation needs. Further research by the project considers UKCP18 model biases in multi-variate relationships [20]. For example, at a given location, when the model tends to be warmer than observed, does it also tend to be too dry or too wet? [20] These interrogations are crucial to understanding uncertainties in current model projections of multi-variate climate hazard and risk, as well the development of future model simulations.

The UKCR project 'STORMY-WEATHER' has developed an objective front identification method that is scalable so that it can be applied to the ERA5 dataset, as well as other high-resolution datasets [3]. A dataset of storm types (combinations of cyclones, fronts and thunderstorms) [8, 21, 22] has been produced for ERA5, which allows hazards (e.g. extreme precipitation and winds) and their co-occurring events to be characterised as associated with particular weather systems. Future changes in the hazards can then be determined based on changes in the frequency or characteristics of the storm types, enabling an analysis of plausible worst-case storms for the future.

4 Improved Physical Understanding of Hazards

By making use of new datasets and characterisation of hazards, the physical understanding of these hazards in the present climate can be refined, thereby improving confidence in future projections. UKCR projects explored the urban influence on changes in local weather extremes through UKCP Local and Regional simulations. The local model reproduced the night-time heat island effect more accurately than the regional model due to improved land-surface and urban representation. The two models give different future projections of the urban heat island effect, with very little change projected by the local model [13] associated with differences in the land-surface model and the influence of soil moisture. These results have implications for urban planning and public health.

For flood hazard estimation, hydrological models have typically used input from regional scale models. The UKCP Local simulation data used in catchment scale models indicate larger increases (or smaller decreases) in future peak river flow than previously found in some catchments. Further, a pilot study for Bristol looking at pluvial flooding using the LISFLOOD hydrodynamic model [14] shows that future changes in flood hazard are very different when the model is driven by UKCP Local data compared with precipitation 'uplifts'. This shows the importance of fully capturing changes in space/time rainfall variability and how it interacts with the landscape and demonstrates the need for a national scale follow-on study.

5 Future Hazards

The UKCP Local projections, providing 12 high-resolution convection-permitting simulations of the future, were used as part of the UKCR project 'FUTURE-DRAINAGE' to develop UK-wide rainfall intensity uplifts [9]. These uplifts were applied to 'design rainfall' at particular return periods (e.g. a one in 50 year event), which is used as the industry standard for all UK flood risk studies and assessments. Through the project, quantitative uncertainty estimates of these uplifts have been provided for the first time; they have already been used by the Environment Agency and the Scottish Environment Protection Agency to develop peak rainfall climate change allowances, used for designing and evaluating flood risk management options. The uplifts are also being used by UK water and sewerage companies to help manage and avoid current and future sewer flooding [4]. The UKCP Local projections were also used to develop new time series rainfall projections at very high temporal resolutions (sub-five-minute) for the UK water industry, to better understand and manage impacts on pollution spills from sewer networks [4].

The UKCR project 'Improving Climate Hazard Information' found that, over the past 100 years, extreme precipitation events during autumn increased in frequency by 60% [23] and are expected to increase even more in future [24]. Mean autumn rainfall is expected to decrease in future, due to an increase in dry summer-type weather patterns and fewer wet winter-type patterns, based on the latest Coupled Model Intercomparison Project (CMIP6) models and UKCP climate projections [25]. Complementing this research, the 'FUTURE-STORMS' project also found that summer and autumn precipitation is heavily influenced by organised convective systems on scales of up to 50km, and the contribution of these systems to heavy precipitation is projected to at least double in future [26]. This results in an unequal distribution of heavy precipitation events in time and space. Projections show a 14-fold increase in slow-moving convective storms, with the potential for high precipitation accumulations over land across Europe by 2100 [27]. These results indicate a simultaneous increase in risk from both flood and drought.

Using the UKCP18 ensemble, the UKCR project 'STORMY-WEATHER' predicted significant future increases in the frequency of extreme windstorms over the UK; one in 20-year storms in the 1990s could occur once every 10 years by the 2070s under a high emissions scenario [28]. This is consistent with the projected increase in intensity of the strongest storms found using CMIP6 models [29], and the projected increase in the occurrence of sting jets found using convection-permitting climate models (CPM) [30]. In terms of climate hazards, these windstorms pose the greatest risk to electricity distribution networks in the UK and are a significant cause of insured losses in the sector. Communication with relevant stakeholders on the characterisation of hazards is ongoing, via the following projects: 'Climate services for a Net Zero resilient world' (CS-N0W) funded by the Department for Business, Energy and Industrial Strategy (BEIS) (https://www.gov.uk/government/organisations/department-for-business-energy-and-industrial-strategy); and 'Assessment of climate change event likelihood embedded in risk assessment targeting electricity distribution' (ACCELERATED) funded by Western Power Distribution (now National Grid). Knowledge of the precursors of sting jets has also been used to provide a new forecasting tool at the Met Office, which has already been useful for producing timely weather warnings [31].

The UKCP simulations were further used in the UKCR project 'Multiple Hazards', which looked at how compound hazards may affect the UK agricultural sector, such as the increase in hot dry summers like that of 2018 [7]. When high temperatures couple with high humidity, the risk of heat stress in livestock and blight in potato crops increases; for example, there may be a tenfold increase in the number of days of dairy cattle heat stress in the next 50 years in the South West of England [7]. These results were communicated to industry partners and fed into the

United Kingdom Food Security Report, published by the Department for Environment, Food and Rural Affairs (Defra) in December 2021.

6 Distilling Climate Information

The advancements in our understanding and projection of climate hazards achieved by the UKCR programme are of significant value to the production of user information on climate risk. To understand the impact of climate hazards on UK communities, we must consider how to quantify uncertainties around the fidelity of models, the use of different models, bias adjustment and projections with different warming levels. For example, in the UKCR 'Risk Assessment Frameworks' project, heat-stress risk was estimated using an open-source risk assessment framework (CLIMADA) and temperature and humidity data from multiple data sources. Part of the uncertainty associated with risk can be attributed to hazard information, highlighting the need to further understand the hazards and how best to bias-correct in the data sources, as well as having improved observations against which to evaluate the data sources [32]. Understanding these biases is also key for the multi-variate relationships between heat and humidity used to estimate risk to agriculture and other sectors [7]. Developing an uncertainty budget for risk (a function of hazard, exposure and vulnerability) that includes such nuances would be valuable for decision-making.

According to findings from the 'EuroCORDEX-UK' project, the uncertainty described above is associated with the different levels of warming seen in different model projections. While the UKCP18 and EuroCORDEX ensembles display similar biases in surface air temperature over the UK in the present day, they project very different levels of warming after this, resulting in corresponding differences in hazards such as heat, drought and extreme precipitation. Further understanding of the rates of warming in the simulations—or associating impacts to global warming levels rather than timescales—would help to constrain these projections.

7 Conclusions

The UKCR projects described above highlight the importance of considering the nuances between models when examining how climate-related hazards may change in future in the UK. Uncertainties between the UKCP Regional and Local simulations suggest that it is essential to have a physical understanding of the hazards when interpreting model output.

Incorporating more detailed information into the smaller-scale ensemble would further add confidence in the projections—capturing future urban changes, for example, is important because of the influence of the built environment on convective storms, and because the effect of urbanisation on climate impacts is unclear. Some projects have used data from model simulations at regional (or lower) resolution; this could be extended to the higher resolution convection-permitting model data, which provides a much more reliable representation of hourly rainfall, local extremes and future change. However, improvements to the uncertainty estimates associated with the CPMs could be made by using a multi-model GCM ensemble to drive the CPM [33] or ensembles of CPMs themselves.

New physical (e.g. flooding) and statistical (e.g. heatwave) models have contributed to projections of worst-case scenarios for the hazards in question. In addition, there has been a focus on the physical causes of hazards; for example, precipitation extremes have been considered in terms of slow-moving storms, weather system types and seasonal weather patterns. This information helps with the interpretation of future hazards, which is a function of both the changing frequency and the hazard characteristics of the driver. Evaluation is now needed to assess how the new information (e.g. precipitation uplifts) is actually being used.

A key challenge going forward is how to convert new information on hazard to estimates of risk, by considering the risk budget of hazard, vulnerability and exposure. Data from CPM (and other) models needs to be fed into hydrological (and other) models on a national scale, requiring a much better understanding of vulnerability and exposure and improved collaboration between different disciplines.

REFERENCES

1. Intergovernmental Panel on Climate Change (IPCC). 2021. *Climate Change 2021: The Physical Science Basis. Contribution of Working Group I to the Sixth Assessment Report of the Intergovernmental Panel on Climate Change* [Masson-Delmotte, V., P. Zhai, A. Pirani, S.L. Connors, C. Péan, S. Berger, N. Caud, Y. Chen, L. Goldfarb, M.I. Gomis, M. Huang, K. Leitzell, E. Lonnoy, J.B.R. Matthews, T.K. Maycock, T. Waterfield, O. Yelekçi, R. Yu, and B. Zhou (eds.)]. Cambridge University Press, Cambridge, United Kingdom and New York, NY, USA.
2. Ruane, A.C., Vautard, R., Roshaka Ranasinghe, R., Sillmann, J., Coppola, E., Arnell, N., Faye Cruz, A., Dessai, S., Iles,C.E., Saiful Islam A. K. M., Jones, R.G., Rahimi, M., Carrascal, D.R., Seneviratne,S.I., Servonnat, J., Sörensson, A,A., Sylla, M.B., Tebaldi, C., Wang, W. and Zaaboul, R.

2022. The Climatic Impact-Driver Framework for assessment of risk-relevant climate information. *Earth's Future* **10**(11), e2022EF002803.
3. Sansom, P. G. and Catto, J. L. 2022. Improved objective identification of meteorological fronts: a case study with ERA-Interim. *Geoscientific Model Development. Discuss*, Under review.
4. Dale, M. 2021. Managing the effects of extreme sub-daily rainfall and flash floods - a practitioner's perspective. *Philosophical Transactions of The Royal Society A Mathematical Physical and Engineering Sciences* **379**(2195).
5. Brown, S. J. 2020. Future changes in heatwave severity, duration and frequency due to climate change for the most populous cities. *Weather and Climate Extremes* **30**, 100278.
6. Lyddon, C., Robins, P., Lewis, M., Barkwith, A., Vasilopoulos, G., Haigh, I. and Coulthard, T. 2022. Historic Spatial Patterns of Storm-Driven Compound Events in UK Estuaries. *Estuaries and Coasts* **46**, pp. 30–56.
7. Garry, F. K., Bernie, D. J., Davie, J. C. S. and Pope, E. C. D. 2021. Future climate risk to UK agriculture from compound events. *Climate Risk Management* **32**, 100282.
8. Catto, J., Sansom, P. and Stephenson, D. Temperature scaling of precipitation depends on storm type over Europe. In prep.
9. Chan, S.C., Kendon, E.J., Fowler, H.J., Youngman, B.D., Dale, M. and Short, C. 2023. New extreme rainfall projections for improved climate resilience of urban drainage systems. *Climate Services* **30**, 100375.
10. Shooter, R. and Brown, S. High-resolution estimation of daily precipitation extremes in the United Kingdom using a generalised additive model framework. *Weather and Climate Extremes*. (Under review).
11. Leach, N. J., Watson, P. A. G., Sparrow, S. N., Wallom, D. C. H. and Sexton, D. M. H. 2022. Generating samples of extreme winters to support climate adaptation. *Weather and Climate. Extremes* **36**, 100419.
12. Kendon, E., Short, C., Pope, J., Chan, S., Wilkinson, J., Tucker, S., Bett, P. and Harris, G. 2021 Update to UKCP Local (2.2km) projections. Met Office.
13. Keat, W. J., Kendon, E. J. and Bohnenstengel, S. I. 2021. Climate change over UK cities: the urban influence on extreme temperatures in the UK climate projections. *Climate Dynamics* **57**, pp. 3583–3597.
14. Archer, L., Kendon, E. J. and Al., E. Future change in urban flooding using new convection-permitting climate projections. Water Resource Research, in prep.
15. Harrison, L. M., Coulthard, T. J., Robins, P. E. and Lewis, M. J. 2021. Sensitivity of Estuaries to Compound Flooding. *Estuaries and Coasts* **45**, pp. 1250–1269.
16. Robins, P.E., Lewis, M.J., Elnahrawi, M., Lyddon, C., Dickson, N. and Coulthard, T.J. 2021. Compound Flooding: Dependence at Sub-daily

Scales Between Extreme Storm Surge and Fluvial Flow. *Frontiers in Built Environment* **7**, 116.
17. Youngman, B.D. 2019. Generalized Additive Models for Exceedances of High Thresholds With an Application to Return Level Estimation for U.S. Wind Gusts. *Journal of the American Statistical Association* **114**(528), pp. 1865–1879.
18. Youngman, B. D. 2022. evgam: An R Package for Generalized Additive Extreme Value Models. *Journal of Statistical Software* **103**, pp. 1–26.
19. Economou, T. and Garry, F. 2022. Probabilistic simulation of big climate data for robust quantification of changes in compound hazard events. *Weather and Climate Extremes* **38**.
20. Garry, F. K. and Bernie, D. J. 2023. Characterising temperature and precipitation multi-variate biases in 12 km and 2.2 km UK Climate Projections. *International Journal of Climatology* **43**(6), pp. 2879–2895.
21. Dowdy, A. and Catto, J. L. 2017. Extreme weather caused by concurrent cyclone, front and thunderstorm occurrences. *Scientific Reports* **7**, 40359.
22. Catto, J. L. and Dowdy, A. 2021. Understanding compound hazards from a weather system perspective. *Weather Climate Extremes* **32**.
23. Cotterill, D., Stott, P., Christidis, N. and Kendon, E. 2021. Increase in the frequency of extreme daily precipitation in the United Kingdom in autumn. *Weather and Climate Extremes* **33**.
24. Chan. S.C., Kendon, E.J., Berthou, S., Fosser, G., Lewis, E. and Fowler, H.J. 2020. Europe-wide precipitation projections at convection permitting scale with the Unified Model. *Climate Dynamics* **55**, pp. 409–428.
25. Cotterill, D. F., Pope, J. O. and Stott, P. A. 2022. Future extension of the UK summer and its impact on autumn precipitation. *Climate Dynamics* **1**, pp. 1–14.
26. Chan, S.C., Kendon, E.J., Fowler, H.J., Kahraman, A., Crook, J., Ban, N and Prein, A.F. 2023. Large-scale dynamics moderate impact-relevant changes to organised convective storms. *Communications Earth and Environment* **4**.
27. Kahraman, A., Kendon, E. J., Chan, S. C. and Fowler, H. J. 2021. Quasi-Stationary Intense Rainstorms Spread Across Europe Under Climate Change. *Geophysical Research Letters* **48**, e2020GL092361.
28. Manning, C., Kendon, E. J., Fowler, H. J. and Roberts, N. 2023. Projected increase in windstorm severity and contribution of sting jets over the UK and Ireland. *Weather and Climate Extremes* **40**.
29. Priestley, M. D. K. and Catto, J. L. 2022. Future changes in the extratropical storm tracks and cyclone intensity, wind speed, and structure. *Weather and Climate Dynamics* **3**, pp. 337–360.
30. Manning, C., Kendon, E. J., Fowler, H.J., Roberts, N.M., Berthou, S., Suri, D. and Roberts, M.J.. 2022. Extreme windstorms and sting jets in

convection-permitting climate simulations over Europe. *Climate. Dynamics* **58**, pp. 2387–2404.
31. Gray, S. L., Martínez-Alvarado, O., Ackerley, D. and Suri, D. 2021. Development of a prototype real-time sting-jet precursor tool for forecasters. *Weather* **76**, 369–373.
32. Dawkins, L. C., Bernie, D. J., Lowe, J. A. and Economou, T. 2023. Assessing climate risk using ensembles: A novel framework for applying and extending open-source climate risk assessment platforms. *Climate Risk Management* **40**.
33. Pichelli, E., Coppola, E., Sobolowski, S., Ban, N., Giorgi, F., Stocchi, P., Alias, A., Belušić, D., Berthou, S., Caillaud, C., M Cardoso, R.M., Chan, S., Christensen, O.B., Dobler, A., de Vries, H., Goergen. K., Kendon, E.J., Keuler, K., Lenderink, G., Lorenz, T., Mishra. A.N., Panitz, H-J., Schär, C., Soares, P.M.M., Truhetz, H. and Vergara-Temprado. J. 2021. The first multi-model ensemble of regional climate simulations at kilometer-scale resolution part 2: historical and future simulations of precipitation. *Climate Dynamics* **56**(5), pp. 3581–3602.

Open Access This chapter is licensed under the terms of the Creative Commons Attribution 4.0 International License (http://creativecommons.org/licenses/by/4.0/), which permits use, sharing, adaptation, distribution and reproduction in any medium or format, as long as you give appropriate credit to the original author(s) and the source, provide a link to the Creative Commons license and indicate if changes were made.

The images or other third party material in this chapter are included in the chapter's Creative Commons license, unless indicated otherwise in a credit line to the material. If material is not included in the chapter's Creative Commons license and your intended use is not permitted by statutory regulation or exceeds the permitted use, you will need to obtain permission directly from the copyright holder.

CHAPTER 10

Future Changes in Indicators of Climate Hazard and Resource in the UK

Nigel Arnell, Stephen Dorling, Hayley Fowler, Helen Hanlon, Katie Jenkins and Alan Kennedy-Asser

Abstract

- The UK Climate Resilience Programme (UKCR) funded several projects that have calculated how climate change is likely to affect climate hazards and resources across the UK, using the latest UK Climate Projections (UKCP18).
- Under a high emissions scenario, heatwaves and high temperature extremes become more frequent across the UK, leading to an

Lead Author: Nigel Arnell

Contributing Authors: Stephen Dorling, Hayley J. Fowler, Helen Hanlon, Katie Jenkins, Alan L. Kennedy-Asser

N. Arnell (✉)
University of Reading, Reading, UK
e-mail: n.w.arnell@reading.ac.uk

© The Author(s) 2024
S. Dessai et al. (eds.), *Quantifying Climate Risk and Building Resilience in the UK*,
https://doi.org/10.1007/978-3-031-39729-5_10

increase in human mortality, animal heat stress, potato blight, wildfire danger and damage to road and rail infrastructure. Cold weather extremes continue to occur but become less frequent.
- Also under high emissions, the growing season starts earlier, lasts longer and is warmer; this is particularly beneficial for grassland and viticulture, but the chance of summer drought and dry soils increases. The precise effects vary across different agricultural systems.
- With respect to rainfall, high hourly and daily totals become more frequent, leading to a greater chance of flash flooding. River floods become more frequent in the north and west of the UK, but low river flows and droughts also become more frequent, and water quality in upland water sources declines. The actual size of the change in risk is uncertain, primarily due to uncertainty in exactly how rainfall will change.
- There are large differences in change in risk across the UK. However, the actual size of the change in risk is uncertain, primarily due to uncertainty in exactly how rainfall will change.

Keywords Climate risk · Climate hazards · Climate resources · UKCP18

1 Introduction

Over the last 30 years, several studies have sought to quantify the potential impacts of climate change in the UK in order to understand how risks might change and to inform adaptation and resilience policy in

---

S. Dorling · K. Jenkins
University of East Anglia, Norwich, UK

H. Fowler
Newcastle University, Newcastle Upon Tyne, UK

H. Hanlon
Met Office, Exeter, UK

A. Kennedy-Asser
University of Bristol, Bristol, UK

specific areas. Since 1998, a series of national climate projections have been produced, with the most recent UK Climate Projections published in 2018 (UKCP18) [1]. The UK Climate Resilience Programme (UKCR) funded a series of projects that characterised future climate risks and opportunities across UK sectors under its 'Climate Risks' theme. This chapter draws together published results from these projects. It focuses on indicators of change in climate hazard and resource and concentrates on how these indicators change in a world with continued high emissions. The indicators characterise change in vulnerability and risk, and therefore, help inform the development of policy and decision-making on adaptation and resilience.

The paper is organised into three sections, outlining the potential changes in hazard and resource that have been calculated during the UKCR programme, summarising how the results are being used and identifying important remaining gaps.

2 Changes in Hazard and Resource Across the UK

The weather and climate at a place constitute both a hazard (something that has the potential to cause harm, such as drought and flooding) and a resource (something that enables or constrains an activity, such as renewable energy generation). Heatwaves, floods and droughts are the most obvious hazards, while agriculture and building heating requirements are strongly influenced by climate resources. A change in climate alters these hazards and resources and will impact many areas, including health, infrastructure and the natural environment. Future impacts will depend on how the economy and society changes over time—and what measures are put in place to adapt and increase resilience—so are sensitive to assumptions about both trends and decisions. Another way of looking at the effects of climate change is to examine how indicators that relate to decisions and actions might change: how often, for example, would the thresholds that trigger emergency plans, or that are known to cause operational challenges, be crossed? Such information helps inform the development of adaptation and resilience policy.

Several UKCR projects took this approach; the results are summarised here, grouped into three sectors as classified in the Technical Report for the Third Climate Change Risk Assessment (CCRA3) [2]: (1) natural environment and assets; (2) infrastructure; and (3) health, communities and the built environment. Table 1 presents a high-level overview of changes in hazard and resource, together with some example quantifications for the 2050s, while the following sections provide more detail. It is important to emphasise that the indicators presented do not necessarily characterise the full range of climate risks in the UK, and that the summary concentrates on research undertaken by UKCR projects specifically.

The indicators referred to in this chapter are listed in Table 1, alongside the corresponding UKCR projects. These studies used the latest UKCP18 climate projections [1], so the following section outlines these projections and details how they have been used.

2.1 Climate Projections and the Construction of Climate Scenarios

The UKCP18 land climate projections consist of four strands [1, 15], three of which (global, regional and local) describe changes in climate with the very high emissions 'RCP8.5' scenario, where global average temperature increases well above 4°C by the end of the twenty-first century. The three strands are based on climate models at different spatial resolutions, so provide information at different scales. The finer resolution models typically provide better information on short-duration, localised weather processes. All are based on variants of the Met Office HadGEM3.05 climate model, and therefore represent only a portion of the range of possible modelled future climates. The global strand combines the HadGEM3.05 model with other climate models, so gives a broader range of possible outcomes.

The fourth strand—the probabilistic projections—not only includes many more climate models, but also includes projections for a wider range of assumptions about future emissions. The projections are intended to characterise the most complete range of uncertainty in how climate is likely to change in the UK in future. In practice, most of the studies of the UKCR programme have used the regional and local strand HadGEM RCP8.5 projections because they are spatially and temporally coherent; very few have used the probabilistic projections. Different studies have used slightly different time periods to define the current reference period and have used a variety of approaches (bias-adjustment or delta methods) to create climate scenarios.

Table 1 Changes in indicators of climate hazard and resource by the 2050s, based on the central estimate from the HadGEM UKCP18 strand (global, regional or local), with very high RCP8.5 emissions

Indicator	Projected change	UKCR project
Natural environment and assets		
The crop growing season starts earlier and lasts longer	• growing season starts around 30 days earlier and ends around 15 days later	Climate Risk Indicators Ref: uk-cri.org
Growing degree days increase	• growing degree days increase by over 50%, with slightly greater increases in the north	Climate Risk Indicators Ref: uk-cri.org
Frost days decrease	• frost days decrease by around 60%	Climate Risk Indicators Ref: uk-cri.org
Growing seasons for viticulture become warmer and longer	• growing season temperatures increase by around 1.5°C and growing degree days increase by 25% (by 2021–2040) • early season air frosts occur on around five fewer days per year • high inter-annual variability remains a consistent feature	CREWS-UK Ref: Nesbit et al. [3]
Soil moisture deficits increase in summer	• average potential soil moisture deficit increases by 75% in England and over 50% in Scotland	Climate Risk Indicators Ref: uk-cri.org
Agricultural drought risk increases	• average proportion of time *in rainfall* drought nearly doubles in England, and when evaporation is included increases by a factor of four • increases are lower in Scotland and Wales	Climate Risk Indicators Ref: uk-cri.org
Warm and dry summers become more common	• average chance of getting two warm and dry months increases from 10 to 40%	Multiple Hazards Ref: Garry et al. [4]

(continued)

Table 1 (continued)

Indicator	Projected change	UKCR project
Potato blight becomes more frequent	• number of days with blight risk increases by 24% in eastern England and 67% in eastern Scotland (the two main potato growing regions)	Multiple Hazards Ref: Garry et al. [4]
Thermal heat stress to dairy cattle increases	• number of days with stressful conditions increases by a factor of 24 in southwest England	Climate Risk Indicators Ref: uk-cri.org
Wildfire danger increases across the UK	• chance of days with 'very high' wildfire danger more than doubles across England, Wales and Northern Ireland, with slightly lower increases in Scotland	Climate Risk Indicators Ref: Arnell et al. [5] and Perry et al. [6]
Infrastructure		
Road accident risk due to ice decreases	• average number of days across UK with risk of icing falls by 70%	Climate Risk Indicators Ref: Arnell et al. [7]
High temperature extremes affecting rail infrastructure increase	• average number of days in England with maximum temperatures greater than 26°C increases by factor of five, and by a greater amount in Scotland and Wales	Climate Risk Indicators Ref: Arnell et al. [7]
Adverse rail operating days increase in England, but decrease in Scotland	• adverse days nearly double in England • adverse days decrease by 25% in Scotland but increase after 2050	Climate Risk Indicators Ref: Arnell et al. [7]

(continued)

Table 1 (continued)

Indicator	Projected change	UKCR project
Short-duration rainfalls become more frequent	• average number of days with high rainfall in England and Wales doubles • in the north and west the 30-year return period is over 30% larger than at present, with smaller increases in the south and east	FUTURE-DRAINAGE Ref: Chan et al. [8] Multiple Hazards Ref: Hanlon et al. [9]
River flows decrease in summer and (in the north and west) increase in winter	• average winter runoff in Wales increases by 11% • average summer runoff falls by over 40% in England and Wales	Climate Risk Indicators Ref: Arnell et al. [7]
River floods become larger and more frequent in the north and west	• the 10-year flood is at least 10% larger in the north and west • the current 10-year flood occurs up to twice as often in the north and west	Climate Risk Indicators Ref: Kay et al. [10]
Low river flows decrease and become more frequent across the UK	• the 10-year return period low river flow is halved across England and Wales, and reduced by a third in Scotland • the current 10-year return period low flow occurs four times as often in England and Wales, and twice as often in Scotland	Climate Risk Indicators Ref: Kay et al. [10]
River drought frequency increases across the UK	• the amount of time in severe hydrological drought doubles across Britain	Climate Risk Indicators Ref: Arnell et al. [7]
Dissolved organic matter increases in upland drinking water sources	• dissolved organic matter concentrations increase by over 30% in autumn in example upland catchments	FREEDOM-BCCR Ref: Monteith et al. [11]

(continued)

Table 1 (continued)

Indicator	Projected change	UKCR project
Health, communities and the built environment		
Heatwaves and heat-health alerts become much more frequent, particularly in the south and east	• average chance of a Met Office defined heatwave increases from 42 to 96% in England • average chance of a heat-health alert increases from 7 to 63%	Climate Risk Indicators Ref: Arnell and Freeman [12]
High temperature extremes become much more frequent, particularly in the south and east	• average number of days per year with maximum temperatures greater than 25°C increases from eight to 25 across the UK • average number of tropical nights per year (minimum temperatures greater than 20°C) increases from less than 0.02 to almost three in England • average number of days per year with heat stress (wet-bulb globe temperature in shade > 25°C) increases from 0.1 to over four in England	Climate Risk Indicators Ref: Arnell and Freeman [12] uk-cri.org OpenCLIM Ref: Kennedy-Asser et al. [13]
Heat-related mortality increases	• heat-related mortality in summer increases by over 30%	Health Sector Resilience Ref: Huang et al. [14]
Cold weather extremes reduce but remain common	• average chance of a cold weather alert decreases from almost 100% to 56% in England	Climate Risk Indicators Ref: Arnell and Freeman [12]
Cold-related mortality decreases	• cold-related mortality in winter decreases by 18%	Health Sector Resilience Ref: Huang et al. [14]

(continued)

Table 1 (continued)

Indicator	Projected change	UKCR project
Heating degree days reduce	• heating degree days decrease by 30% in England and 48% in Scotland	Climate Risk Indicators Ref: Arnell et al. [7] Multiple Hazards Ref: Hanlon et al. [9]
Cooling degree days increase	• cooling degree days increase by a factor of four in England and five in Scotland	Climate Risk Indicators Ref: Arnell et al. [7] Multiple Hazards Ref: Hanlon et al. [9]

2.2 Natural Environment and Assets

While this sector encompasses natural environments, agriculture, forestry and the landscape [16], studies published thus far have concentrated on wildfire (the hazard) and on indicators relevant to agriculture.

Wildfires in the UK are typically a result of human action (usually inadvertent), but the fire danger at a place depends on current and preceding weather. Projected higher temperatures, lower humidity and increased drought conditions result in increased wildfire danger across the UK, particularly in the south and east [5, 6]. Each indicator presents a different degree of change with respect to wildfire danger, but all point to an increase in risk.

Regarding UK agriculture, projected warmer temperatures increase the duration of the growing season, which serves to both increase growing degree days and reduce early season frost frequency—conditions that benefit grass and potentially other crops such as vines [3, 4, 9, 17]. However, summer soil moisture deficits increase across the UK (particularly in the south and east) due to drier summers and greater evaporation. This potentially limits production of crops without supplementary irrigation. Agricultural drought risk increases with increasing temperature across the UK, with a large uncertainty range primarily due to uncertainty in projected changes in rainfall [9, 17]. With very high RCP8.5 emissions, the chance of having two warm and dry summer months in a year—which is challenging to agriculture—would increase from less than 10% now to over 40% by the 2050s, but again with a large uncertainty range [4].

Projected high temperatures, coupled with high humidity, also lead to increased risk of potato blight (particularly in typically cool areas)

[4], lower milk yields from dairy cattle owing to increased heat stress (particularly in the south and east) [4, 17] and a substantial increase in the occurrence of debilitating sheep parasites (in the south and west) [17]. The annual number of days with very wet soils, which limits access to land, decreases in future [17]. This reduction occurs mainly during autumn, which would potentially help with the drilling of winter crops and establishment of robust rooting systems.

Taken together, these changes in climate imply some opportunities for UK agriculture (especially at lower emission scenarios), but also mean increasingly challenging conditions associated with extreme events for many types of farming. The precise impacts on yields will vary between crop type and variety, depending on their resilience to changes in temperature, rainfall and associated pests and diseases.

2.3 Infrastructure

This sector covers the UK infrastructure providing services such as heating, lighting, mobility, freshwater and sanitation to society and protecting against extreme events [18].

The number of days with 'heavy' rainfall above thresholds used in the Met Office National Severe Weather Warning Service is projected to increase into the future [9], implying more frequent flash flooding. The magnitude of rainfall events with a specific chance of occurring—such as one in 30 years—is projected to increase [8, 19]. The increases are proportionally greater in the north and west.

Changes in hourly and daily extreme rainfall such as these would directly affect flash flood risk (e.g. in urban areas), but not necessarily directly affect river flood risk. Small and impermeable catchments respond rapidly to short-duration rainfall, while flooding in larger catchments or catchments with more storage reflects longer-term accumulations of rainfall over days or weeks. The general picture is for an increase in flood risk (interpreted as change in either magnitude or frequency) across northern and western parts of the UK, resulting from increased rainfall and potentially slower-moving rainstorms [20]. In the south and east, there is more of a mixed picture [10, 21], with considerable uncertainty in the amount of change. The projected increase in flooding increases the erosion risk to bridges and other infrastructure [22].

At the same time, reductions in spring, summer and autumn rainfall across southern and eastern England result in lower summer and autumn river flows [10, 21], increased frequencies of hydrological drought [23] and associated pressure on water resources and the water environment—although the uncertainty range is again large.

High temperature extremes affect the performance and maintenance of road and rail infrastructure (e.g. causing road surfaces to melt and track/signalling equipment to malfunction), while low temperatures increase road accident risk through road icing [23]. High temperature extremes are projected to become much more frequent, reducing the reliability of road and rail infrastructure and increasing the frequency of failure. Low temperature extremes become less frequent but will continue to occur.

Adverse weather conditions can also affect operations on the railways. Characterised as hot, cold, wet and windy extremes, occurrences of 'adverse weather' impacts punctuality standards. Across England and Wales, the number of days with adverse weather is projected to increase with warming [23], primarily because hot days increasingly dominate. In contrast, adverse conditions in Scotland are strongly influenced by cold weather, and these decrease over time.

The reliability of renewable energy supplies is potentially affected by changes in climate, but changing demand (see below) will be more important than the projected small variation in the frequency of 'wind drought' [24, 25].

2.4 Health, Communities and the Built Environment

This sector concentrates on risks to the UK population, focusing on health and wellbeing, as well as the built environment [26].

Extreme high temperatures in the UK are projected to increase more rapidly than global average temperatures [13, 27]; the number of days above specific temperature thresholds increases substantially, particularly in the south and east [9, 23]. Two definitions of heatwave are currently used operationally, one by the Met Office to declare a heatwave (primarily for communications purposes) and one used in England in the health and social care system. Under both definitions, the number of events is projected to increase significantly. When translated into risk of human mortality [14, 28], risks increase with temperature in a very non-linear way and accelerate as temperature rises. Extreme heat stress arises where high temperatures are associated with high humidity; high heat-stress days are currently very rare in the UK, but the chance of experiencing them in future increases very substantially in the south-east [12, 13]. At the other extreme, cold weather events become less frequent [9, 12], but will still occur with sufficient frequency that they need to be considered for planning purposes.

Extreme windstorms are likely to increase in frequency throughout the twenty-first century, with a large proportion likely to feature very high windspeeds due to 'sting jets' [29] leading to an increased risk of property damage.

Heating degree days (a proxy for the demand for residential heating) decrease by about 18% and 35% at 2°C and 4°C warming, respectively [9, 23], with the percentage change relatively consistent across the UK. The uncertainty range is small because there is relatively little uncertainty in how UK average temperature changes with global average temperature. Cooling degree days (a proxy for overheating and therefore cooling requirements) increase from a low baseline, and therefore, percentage changes are potentially misleading. However, there is much more variability across the UK, with the greatest increase experienced in the south and east.

3 How Have the Results Been Used so Far?

A wide range of indicators of current and future climate hazard and resource across the UK are available from the Climate Risk Indicators website, an interactive tool enabling users to map indicators and plot time series at scales ranging from local authority area to the four nations of the UK. This resource has been used by several organisations to understand local and regional climate risks, and some (e.g. The Wildlife Trusts) [30] have combined the indicators with their own data to create customised risk maps. Results and figures also figure prominently in the CCRA3 Technical and Synthesis Reports. The UK Heat Stress Vulnerability website allows users to produce maps of current and future heat stress, combining metrics of heat hazard with metrics of vulnerability, which has been used by the Welsh Government [31]. The projected changes in short-duration rainfall produced in the UKCR project 'FUTURE-DRAINAGE' [18] have been used to define new peak rainfall climate change allowances for both England [32] and Scotland [33]. The UK Government's CCRA3 [34] refers to both the climate risk indicators website and 'FUTURE-DRAINAGE'. The results from several projects have also been used as the basis for more specific investigations for individual public and private sector organisations.

4 Gaps and Challenges

Taken together, the UKCR-funded studies of changing climate risk in the UK have demonstrated the potential for large, adverse impacts and increases in risk. Also highlighted, however, are the large uncertainty ranges, primarily due to uncertainty in how mean and extreme rainfall across the UK will change over time. The studies have used slightly different approaches in detail, but all produce comparable results. There are, however, three main gaps.

First, the studies summarised here have typically focused on specific sectors or physical hazards on the land. While these were all assumed to be important, they were not selected on the basis of their relative contributions to climate change risk in the UK and do not provide a comprehensive coverage. For example, no studies have yet published indicators relating to potential risks to the natural environment (except due to wildfire hazard).

Second, most of the assessments so far have focused on the regional and local strands of the UKCP18 projections, which assume a very high rate of increase in future emissions [35, 36]. It is difficult to link these directly to potential changes under lower emissions, following the Climate change Committee's guidance [37] to adapt to a two degree world but prepare for four degrees—although some analysis has been undertaken using subsets of the probabilistic strand [7] or presented impacts by level of warming [9, 13, 23]. In practice, there is little difference between emissions scenarios to the 2040s, compared with the uncertainty range, so the gap is more relevant over the longer term. Also, the regional and local strands used in most studies are based on a climate model with a relatively high climate sensitivity. The use of just a subset of the full suite of UKCP18 projections means that uncertainty ranges are probably underestimated.

Third, with the exception of the agriculturally focused 2021 study by Garry and others [4], there have been few assessments of how compound events and extremes might change in future (e.g. the cumulative impact of wind, rain and storm surges, or hot, dry and high fire danger conditions).

5 Conclusions

The studies summarised here together enable a consistent assessment of potential changes in climate hazards and resources across the UK using the same underlying climate scenarios, in terms directly relevant to the management of climate risks. Future risk and future impacts depend on how these changes in hazard and resource interact with changes in exposure and vulnerability.

References

1. Lowe, J.A., Bernie, D., Bett, P., Bricheno, L., Brown, S., Calvert, D., Clark, R., Eagle, K., Edwards, T., Fosser, G., Fung, F., Gohar, L., Good, P., Gregory, J., Harris, G., Howard, T., Kaye, N., Kendon, E., Krijnen, J., Maisey, P., McDonald, R., McInnes, R., McSweeney, C., Mitchell, J.F.B., Murphy, J., Palmer, M., Roberts, C., Rostron, J., Sexton, D., Thornton, H., Tinker, J., Tucker, S., Yamazaki, K., and Belcher, S. 2018. UKCP18 Science Overview Report. [Online]. Available at: UKCP18-Overview-report.pdf (metoffice.gov.uk).
2. Betts, R.A., Haward, A.B. and Pearson, K.V. (eds) 2021. *The Third UK Climate Change Risk Assessment Technical Report*. [Online]. Available at: Technical-Report-The-Third-Climate-Change-Risk-Assessment.pdf (ukclimaterisk.org).
3. Nesbitt, A., Dorling, S., Jones, R., Smith, D.K.E., Krumins, M. Gannon, K.E. and Conway, D. 2022. Climate change projections for UK viticulture to 2040: a focus on improving suitability for Pinot Noir. *Oeno One* **56** (3), pp. 69–87.
4. Garry, F.K., Bernie, D.J., Davie, J.C.S. and Pope, E.C.D. 2021. Future climate risk to UK agriculture from compound events. *Climate Risk Management* **32**, 100282.
5. Arnell, N.W., Freeman, A. and Gazzard, R. 2021. The effect of climate change on indicators of fire danger in the UK. *Environmental Research Letters* **16**, 044027.
6. Perry, M., Vanvyve, E., Betts, R.A. and Palin, E.J. 2022. Past and future trends in fire weather for the UK. *Natural Hazards and Earth System Sciences* **22**, pp. 559–575.
7. Arnell, N.W., Kay, A.L., Freeman, A., Rudd, A.C. and Lowe. J.A. 2021. Changing climate risk in the UK: a multi-sectoral analysis using policy relevant indicators. *Climate Risk Management* **31**, 100265.
8. Chan, S.C., Dale, M., Fowler, H.J. and Kendon, E.J. 2021. Extreme precipitation return level changes at 1, 3, 6, 12, 24 hours for 2050 and 2070,

derived from UKCP Local Projections on a 5km grid for the FUTURE-DRAINAGE Project. [Online]. Available at: https://catalogue.ceda.ac.uk/uuid/18f83caf9bdf4cb4803484d8dce19eef.
9. Hanlon, H.M., Bernie, D., Carigi, G. and Lowe, J.A. 2021. Future changes to high impact weather in the UK. *Climatic Change* **166**, 50.
10. Kay, A.L., Griffin, A., Rudd, A.C., Chapman, R.M., Bell, V.A. and Arnell, N.W. 2021. Climate change effects on indicators of high and low river flows across Great Britain. *Advances in Water Resource*, **151**, 103909.
11. UK Centre for Ecology and Hydrology. 2021. How will climate change influence levels of dissolved organic matter in upland drinking water sources? FREEDOM-BCCR briefing note 5 to the water industry. [Online]. Available at: FREEDOM_BCCR_Climate_Effects_Modelling_05.pdf (ceh.ac.uk).
12. Arnell, N.W. and Freeman, A. 2021. The impact of climate change on policy-relevant indicators of temperature extremes in the United Kingdom. *Climate Resilience and Sustainability* **1**(2), e12.
13. Kennedy-Asser, A.T., Owen, G., Griffith, G.J., Andrews, O., Lo, Y.T.E., Mitchell, D.M., Jenkins, K. and Warren, R.F. 2022. Projected risks associated with heat stress in the UK Climate Projections (UKCP18). *Environmental Research Letters* **17**(3), 034024.
14. Huang, W.T.K., Braithwaite, I., Charlton-Perez, A., Sarran, C. and Sun, T. 2022. Non-Linear Response to Global Climate Change of Temperature-Related Mortality Risk in England and Wales. *Environmental Research Letter*, **17**, 034017.
15. Murphy, J., Harris, G., Sexton, D., Kendon, L., Bett, P., Clark, R., Eagle, K., Fosser, G., Fung, F., Lowe, J., McDonald, R., McInnes, R., McSweeney, C., Mitchell, J., Rostron, J., Thornton, H., Tucker, S. and Yamazaki, K. 2019. *UKCP18 Land Projections: Science Report*. [Online]. Available at: PowerPoint Presentation (rmets.org).
16. Berry, P. and Brown, I. 2021. Natural environment and assets. In: *The Third UK Climate Change Risk Assessment Technical Report* (eds. Betts, R.A., Haward, A.B. and Pearson, K.V.). [Online]. Available at: CCRA3-Chapter-3-FINAL.pdf (ukclimaterisk.org).
17. Arnell, N.W and Freeman, A. 2021. The effect of climate change on agro-climatic indicators in the UK. *Climatic Change* **165**, 40.
18. Jaroszweski, D., Wood, R. and Chapman, L. 2021. Infrastructure. In: *The Third UK Climate Change Risk Assessment Technical Report* (eds. Betts, R.A., Haward, A.B. and Pearson, K.V.). [Online]. Available at: CCRA3-Chapter-4-FINAL.pdf (ukclimaterisk.org).
19. Dale, M. 2021. Future Drainage. Guidance for water and sewerage companies and flood risk management authorities: recommended uplifts for applying to design storms. [Online]. Available at: (FUTURE_DRAINAGE_Guidance_for_applying_rainfall_uplifts.pdf (ceda.ac.uk)).

20. Kahraman, A., Kendon, K.J., Chan, S.C. and Fowler, H.J. 2021. Quasi-stationary intense rainstorms spread across Europe under climate change. *Geophysical Research Letters* **48**(13), e2020GL092361.
21. Kay, A.L., Rudd, A.C., Fry, M., Nash, G., and Allen, S. 2021. Climate change impacts on peak river flows: combining national-scale hydrological modelling and probabilistic projections. *Climate Risk Management* **31**, 100263.
22. Li, X., Cooper, J.R. and Plater, A.J. 2021. Quantifying erosion hazards and economic damage to critical infrastructure in river catchments: impact of a warming climate. *Climate Risk Management* **32**, 100287.
23. Arnell, N.W., Freeman, A., Kay, A.L., Rudd, A.C. and Lowe, J.A. 2021. Indicators of climate risk in the UK at different levels of warming. *Environmental Research Communications* **3**, 095005.
24. Dawkins, L. *et al*. 2021. *Adverse Weather Scenarios for Future Electricity Systems: Developing the dataset of long-duration events*. Met Office.
25. Dawkins, L. Rushby, I., Pearce, M., Wallace, E. and Butcher, T. 2021. Adverse Weather Scenarios for Future Electricity Systems. [Online]. Available at: https://catalogue.ceda.ac.uk/uuid/7beeed0bc7fa41feb10be22ee9d10f00.
26. Kovats, S. and Brisley, R. 2021. Health, communities and the built environment. In: *The Third UK Climate Change Risk Assessment Technical Report* (eds. Betts, R.A., Haward, A.B. and Pearson, K.V.). [Online]. Available at: CCRA3-Chapter-5-FINAL.pdf (ukclimaterisk.org).
27. Kennedy-Asser, A.T, Andrews, O., Mitchell, D.M. and Warren, R.F. 2020. Evaluating heat extremes in the UK Climate Projections (UKCP18). *Environmental Research Letters* **16**, 014039.
28. Ibbetson, A., Milojevic, A., Mavrogianni, A., Oikonomou, E., Jain, N., Tsoulou, I., Petrou, G., Gupta, R., Davies, M. and Wilkinson, P. 2021. Mortality benefit of building adaptations to protect care home residents against heat risks in the context of uncertainty over loss of life expectancy from heat. *Climate Risk Management* **32**, 100307.
29. Manning, C., Kendon, E.J., Fowler, H.J., Roberts, N.M., Berthou, S., Suri, D. and Roberts, J.M. 2022. Extreme windstorms and sting jets in convection-permitting climate simulations over Europe. *Climate Dynamics* **58**, pp. 2387–2404.
30. Wildlife Trusts. 2022. Changing Nature: A Climate Adaptation Report by the Wildlife Trusts. [Online]. Available at: AdaptationReport.pdf (wildlifetrusts.org).
31. Welsh Government. 2021. Explanatory Memorandum to the Climate Change (Wales) Regulations 2021. [Online]. Available at: EM template for sub leg (senedd.wales).

32. Environment Agency. 2022. Flood risk assessments: climate change allowances. [Online]. Available at: https://www.gov.uk/guidance/flood-risk-assessments-climate-change-allowances.
33. Scottish Environment Protection Agency. 2022. *Climate change allowances for flood risk assessment in land use planning.* Volume 2 LUPS-CC1 (No longer available online).
34. HM Government. 2022. Climate Change Risk Assessment. [Online]. Available at: UK Climate Change Risk Assessment 2022 - GOV.UK (www.gov.uk).
35. Hausfather, Z. and Peters, G.P. 2020. Emissions – the 'business as usual' story is misleading. *Nature* 577, pp. 618–620.
36. Schwalm, C.R., Glendon, S. and Duffy, P.B. 2020. RCP8.5 tracks cumulative CO_2 emissions. *PNAS* 117(33), pp. 19656–19657.
37. Climate Change Committee. 2021. *Independent Assessment of UK Climate Risk. Advice to Government for the UK's third Climate Change Risk Assessment (CCRA3)*. [Online]. Available at: Independent Assessment of UK Climate Risk - Climate Change Committee (theccc.org.uk).
38. Ronasinghe, R. et al. 2021. Climate change information for regional impact and for risk assessment. In: IPCC Climate Change 2021: The Physical Science Basis. Contribution of Working Group I to the Sixth Assessment Report of the Intergovernmental Panel on Climate Change (eds. Masson-Delmotte, V. et al). [Online]. Available at: Chapter 12: Climate Change Information for Regional Impact and for Risk Assessment | Climate Change 2021: The Physical Science Basis (ipcc.ch).

Open Access This chapter is licensed under the terms of the Creative Commons Attribution 4.0 International License (http://creativecommons.org/licenses/by/4.0/), which permits use, sharing, adaptation, distribution and reproduction in any medium or format, as long as you give appropriate credit to the original author(s) and the source, provide a link to the Creative Commons license and indicate if changes were made.

The images or other third party material in this chapter are included in the chapter's Creative Commons license, unless indicated otherwise in a credit line to the material. If material is not included in the chapter's Creative Commons license and your intended use is not permitted by statutory regulation or exceeds the permitted use, you will need to obtain permission directly from the copyright holder.

CHAPTER 11

What Has Been Learned About Converting Climate Hazard Data to Climate Risk Information?

Dan Bernie, Freya Garry, Katie Jenkins, Nigel Arnell, Laura Dawkins, Alistair Ford, Alan Kennedy-Asser, Paul O'Hare, Rachel Perks, Victoria Ramsey and Paul Sayers

Abstract

- Understanding climate risks requires consideration of the hazard, vulnerability and exposure.

Lead Authors: Dan Bernie, Freya Garry & Katie Jenkins

Contributing Authors: Nigel Arnell, Laura Dawkins, Alistair Ford, Alan Kennedy-Asser, Paul O'Hare, Rachel Perks, Victoria Ramsey & Paul Sayers

D. Bernie (✉) · F. Garry (✉) · L. Dawkins · R. Perks · V. Ramsey
Met Office, Exeter, UK
e-mail: dan.bernie@metoffice.gov.uk

A. Kennedy-Asser
University of Bristol, Bristol, UK

© The Author(s) 2024
S. Dessai et al. (eds.), *Quantifying Climate Risk and Building Resilience in the UK*,
https://doi.org/10.1007/978-3-031-39729-5_11

- The understanding and quantification of climate vulnerabilities is central to developing valuable assessments of future risks, with close communication between stakeholders and researchers crucial to achieving this.
- Access to existing exposure and vulnerability data is highly fragmented; a centralised authoritative repository, where such data could be combined with climate data, would widen access and facilitate research.
- There is an ongoing need for multiple risk frameworks and tools to address the breadth of climate resilience issues.
- The analysis of compound, cascading and systemic risks would benefit from more focus in the context of national scale risk assessments.

Keywords Climate · Hazards · Risks · Vulnerability · Exposure

1 Introduction

The link between human-induced global warming and changing weather and climate is well documented [1]. Changes to UK climate have been observed over recent decades, with implications for both current and future climate hazards [2]. Climate variability and change, including

---

K. Jenkins (✉) · P. Sayers
University of East Anglia, Norwich, UK

N. Arnell
University of Reading, Reading, UK

A. Ford
Newcastle University, Newcastle upon Tyne, UK

P. O'Hare
Manchester Metropolitan University, Manchester, UK

P. Sayers
Sayers and Partners, Watlington, UK

changes in the severity, frequency and spatial patterns of extreme weather, can have wide ranging impacts on society, the economy and the environment. Examples of impacts include risks to human health due to increased exposure to heat in buildings, risks to people and the economy from climate-related disruption of power systems and risks to soil health and agriculture from increased flooding and drought [3]. Climate risk is commonly defined as a combination of the climate hazard (see also Chapters 9 and 10), exposure and vulnerability, with response sometimes also considered as a separate determinant (Fig. 1).

Interactions between sectors and systems will also affect risk. Clearly defining, representing and combining elements of these components, which can stretch across social, economic and environmental domains, sometimes in an interrelated fashion, is extremely challenging. In addition, most risk assessments do not consider the potential for compound

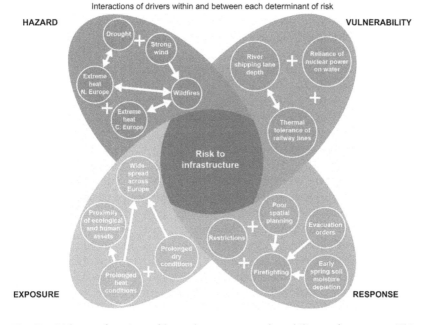

Fig. 1 Risk as a function of hazard, exposure, vulnerability and response. This example illustrates some of the complex interactions that generated risk to infrastructure during the 2018 European heatwave [4]

or cascading consequences [5], which can lead to an underestimate of risk [4]. Different approaches and methodologies to convert climate hazard data to climate risk information have been pursued and applied through the UK Climate Resilience Programme (UKCR).

2 Progress in Climate Risk Quantification—Overview

Qualitative mapping provides a straightforward approach to draw together pre-existing research and secondary sources to identify hazards and multi-sector risks; for example, to identify multi-sector risks at the city level, as exemplified by the UKCR project 'Manchester Climate Action'. This allows exploration of how risks could evolve in future and a reference point for future research [6], supported by input from key stakeholders and expert-led technical assessments where key risks are identified, or information is incomplete. Similarly, developing a better understanding of historical events (and associated risks) by linking observed reports and datasets with modelled hazard data provides a mechanism to better understand or develop triggers/thresholds that can be used to project the risk of future events occurring—an approach applied by the UKCR project 'Environment Agency Incident Response'. Again, stakeholder engagement or co-production is crucial to support access to novel datasets and exploration of data to support new analyses.

Threshold-based methodologies assess climate-related risks by linking hazard data to the exceedance of a given operational or warning threshold (see also Chapter 10). UKCR projects 'Climate Risk Indicators', 'Meeting Urban User Needs' and 'Hazard to Risk' [7, 8] applied this approach to a wide range of risk-related indicators, including: health and well-being; energy use; transport; agriculture; wildfire; heat stress and hydrological indicators. The UKCR project 'Multiple Hazards' also used this method in its study of compound events [9]. The use of impact-specific thresholds, where discernible, can ensure risk indicators are meaningful to end-users and provide information in relevant and understandable terms at a variety of scales. While thresholds are often based on historical/observed data and are assumed to remain static in future, most analyses could be repeated relatively easily with alternative thresholds if required.

Simulation models can capture more complex relationships between hazard, vulnerability and exposure. For example, implications of heat on care settings, given the vulnerability of specific building characteristics and locations to temperatures, were identified in the UKCR project 'ClimaCare' [10]. Detailed modelling such as this provides new insights into how different components of vulnerability and their relative sensitivity, such as building construction, will affect risk at more localised levels.

Catastrophe (CAT) modelling frameworks, typically used by insurance and financial sectors, model the risks of extreme weather, combining hazard, exposure and vulnerability data. Extensions have allowed future climate risks to be estimated in several UKCR projects; 'AquaCAT' achieved this by enabling the changing spatial structure of flood events to be reflected in national flood risk assessments, whereas the 'Multiple Hazards' project created spatially coherent assessments of heat-related risk and 'Risk Assessment Frameworks' examined heat impacts on physical outdoor work capacity, with risks quantified in terms of cost and days affected.

Systems-based approaches move away from considering individual risks in isolation, aiming to capture the interconnections and interdependencies of risks within a single framework. Bringing together diverse models and methodologies allows multiple sectors to be analysed in a comprehensive and consistent manner. Under 'OpenCLIM', progress has been made in how we design integrated frameworks, develop linkages between models and incorporate adaptation into assessments. However, coupling models also increases the complexity of data requirements, outputs and uncertainties, particularly where multiple dynamics exist.

2.1 Risks and Indicators

Moving from hazard to risk has many challenges, and while the term 'risk' is used universally, often it is sometimes used as shorthand for 'risk-related indicators of exposure or vulnerability only to a climate hazard'. The 'Climate Risk Indicators' project did not explicitly include exposure and vulnerability, although certain indicators were weighted (e.g. based on population) to reflect the hazard and current levels of exposure, and many of the indicators are based on thresholds representing current interpretations of levels of vulnerability. Other studies have mapped overlapping factors that contribute to risk, including data on socioeconomic vulnerability and exposure at the national scale [11]. The UKCR project 'Meeting Urban User Needs' incorporated more localised conditions, drawing together data on vulnerable people, the built environment, green space and council assets. Embedding components such as these will be especially important for decision-makers wanting to understand risk in detail, particularly at smaller scales.

Communicating risk can also be challenging (see also Chapter 12), particularly because the most suitable scale for calculating risk rarely aligns with how risk is best communicated and used. Indeed it is common that the spatial resolution which it is possible to calculate risk at is misaligned with what is needed to inform decisions at different spatial scales [6]. However, the above UKCR projects have demonstrated the benefits of working with stakeholders to maximise utility and uptake—for example, 'Climate Risk Indicators' provided risk indicators based on policy relevant thresholds and critical values, 'Multiple Hazards' provided additional risk-density metrics that allowed a national comparison of results [9], and 'Meeting Urban User Needs' used risk frameworks that align with existing stakeholder frameworks.

3 Areas of Progress in Methodological Development

3.1 Spatially Coherent Event Set Generation Versus Local Return Periods

The spatial characteristics of extreme events are important in assessing the return frequency of a geographically aggregated impact. For example, extreme events may affect multiple assets in a national portfolio, but if return frequencies of events are calculated locally, they do not capture the spatial relationship between impacts on these assets. The risk assessment for a portfolio should be calculated incorporating those spatial relationships to capture the total impacts. This has long been recognised in the insurance sector but should also be considered in assessing systemic or cascading impacts.

The need for considering spatial coherence, and the potential for change in the spatial characteristics, has been examined in UKCR projects 'AquaCAT' [12–14], 'OpenCLIM' [11, 15], and 'Multiple Hazards' [16]. For flooding, 'AquaCAT' predicts an increase in widespread events with very extreme river flows, as well as more widespread events that are formed by much more frequent high levels of river flow. Results from using the tool CLIMADA (https://wcr.ethz.ch/research/climada.html) ('Risk Assessment Frameworks' project), show large increases in the impact of heat on outdoor productivity across the UK, but with potential for regionally differentiated optimal adaptation approaches. Projects such as 'AquaCAT' have advanced novel statistical methods to generate stochastic event sets for both hazard and risk, generating values for the underlying climate simulations.

3.2 Exposure and Vulnerability Data

Vulnerability and exposure can be the key source of uncertainty in risk calculations. It is difficult to fully encompass the range of complex, intersecting factors that these components are contingent upon. For example, data may not exist at the required spatial level, or detailed spatial data may exist but not be available or spatially coherent across different regions of the UK, or projected data may not be available for the desired future time periods.

An important methodological advance is that local, regional and global data underpinning the latest UK Climate Projections (UKCP18) used to model and project hazards, can now be linked to the recently released UK Shared Socioeconomic Pathways (UK-SSPs). The UK-specific SSPs are consistent with the global SSPs, qualitatively and quantitatively describing a set of internally consistent, alternative plausible trajectories of societal development which can be used to support risk assessment. A benefit of this is that climate scenarios can be temporally aligned with projections of socioeconomic change. Certain risks to different sectors, and feedbacks of socioeconomic change, can also be evaluated consistently across a range of socioeconomic futures, as illustrated in 'OpenCLIM', although there are some challenges to using the time-varying UK-SSPs in the context of hazard expressed on global warming levels.

3.3 New Datasets for Hazard, Risk, Vulnerability and Exposure

Throughout the UKCR programme, there have been several developments which have allowed the production and sharing of datasets to better inform assessment of changing climate risk. The availability and use of these by the community will support the evidence base underpinning the next UK Climate Change Risk Assessment. A selection is listed in Table 1.

Table 1 A selection of the new datasets for hazard, vulnerability, exposure and risk developed through the UKCR programme

Class	Project	Dataset
Hazard	Coastal Climate Services	Future storm surges, waves and extreme water levels around the UK coast
Hazard	Risk Assessment Frameworks	Events set of outdoor heat stress
Hazard	EuroCORDEX-UK	Regional climate model (RCM) projections over the UK reformatted to complement the UKCP18 ensemble
Hazard	AquaCAT	AquaCAT flooding event sets
Hazard	Climate Risk Indicators	Risk-informed indicators of climate-related hazards for different UK sectors
Exposure & vulnerability	UK-SSPs	UK-specific socioeconomic pathways (SSPs), down-scaled from the Global/European SSPs
Risk	Meeting Urban User Needs	Heat Vulnerability Index to assess heat risk within the city of Belfast
Risk	Risk Assessment Frameworks	Future of outdoor productivity loss (in person hours) under different socioeconomic and climate futures
Risk	OpenCLIM	Risk-related metrics covering heat stress; inland flooding; risks to water supply; drought; biodiversity and agriculture under different socioeconomic and climate futures
Risk	Multiple Hazards	Maps of future climate risks for cattle heat stress and potato blight occurrence

3.4 Treatment of Uncertainties

There are many sources of uncertainty in the calculation of risk, from the physical characterisation of hazards, the exposure of assets or systems to these hazards and the amount of impact a given hazard will have. These are compounded by the uncertainties around methodological choices in how hazard information is combined with exposure and vulnerability to estimate risk.

Physical uncertainties in climate projections arise from many overlapping factors. Different weather and climate products have been developed over the years which, depending on the intended use, prioritise different types and sources when sampling uncertainty. These uncertainties broadly split between 'aleatoric uncertainty' (the inherent randomness in chaotic systems) and 'epistemic uncertainties', which arise from our incomplete understanding of the physical system and ability to simulate it, including scenario uncertainty arising from the forcing of the system by uncertain human actions [17]. All these sources are considered across different climate products used within UKCR, with some notable advances in the treatment of uncertainty.

The 'EuroCORDEX-UK' project expanded the UKCP18 regional model ensemble with a range of model simulations from EURO-CORDEX [18] to better sample structural uncertainties (from use of different regional and global climate models) as well as the parametric uncertainty (from uncertain physical parameters in a single model) from the original UKCP18 simulations.

An alternative approach was taken by the project 'Coastal Climate Services' which, instead of carrying out new surge and wave simulations, adapted an operational technique for medium- to long-range forecasts to look at the influence of climate change on future coastal risk [19]. Historical wave and storm surge events were linked with North Atlantic pressure patterns and used to quantify the distributions of wave and surge for each pressure pattern. Combining these with projections of future local sea levels and accounting for frequency changes in atmospheric circulation patterns allowed them to assess the changes in coastal risks from extreme water levels.

The approach of using multiple data sources and adapting existing methodologies was taken further in work with CLIMADA, where the uncertainty in future risk was disaggregated with a sensitivity analysis [20]. This served to attribute uncertainty between sources of climate information, methodological choices, assumptions about future socioeconomic trends (from UK-SSP), climate sensitivities[1] and global warming levels. While initially idealised, this combines many of the approaches to dealing with uncertainties that have been used across UKCR.

As well as these specific advances in uncertainty and risk calculation, throughout the UKCR programme different climate products have been used extensively to account for uncertainty. Expert judgement is needed to assess whether a product can credibly represent the hazard of interest and account for uncertainty in the projections, while balancing

---

[1] Climate sensitivity is typically defined as the global temperature rise following a doubling of CO_2 concentration in the atmosphere compared to pre-industrial levels. From: https://www.metoffice.gov.uk/research/climate/understanding-climate/climate-sensitivity-explained#:~:text=Climate%20sensitivity%20is%20typically%20defined,be%20at%20roughly%20520%20ppm.

the computational demands of their use or availability over a specific time period. For example, the Urban Heat Service (an outcome of the 'Meeting Urban User Needs' project) used the highest resolution products available, whereas work on compound hazards affecting UK agriculture ('Multiple Hazards' project) used probabilistic and regional UKCP18 products [9].

4 Gaps and Remaining Challenges

This section represents the views of the authors in terms of their experiences on the balance of opinions held. It is acknowledged that there may be specific sectors or organisations where the remaining challenges differ to the views expressed here.

4.1 Hazards

Availability of, and access to, climate information needed to calculate future risk has improved over recent years, owing to model advances in complexity, horizontal resolution and sampling of uncertainties driven by both international and UK programmes such as UKCR and UKCP18. The continued development of convective scale climate simulations (~1km horizontal resolution) has driven improved understanding of extreme rainfall events in particular.

However, no 'best' set of climate products exists for all use cases—from the user perspective, deciding which tool to use (with limited resources) is challenging, requiring an understanding of the relevant hazards and the characteristics of the different climate products. This is particularly complex where multiple impacts compound the effects of each other, either directly or indirectly, or over different time scales. Closer communication between climate research and impact sectors would help develop a shared understanding of sector vulnerabilities and climate model capabilities, supporting a more insightful application of climate data to resilience issues and ultimately enabling more valuable advice and services.

There is also often a need for calibration or 'bias-correction' of climate data before calculating impact. As with the choice of climate products, deciding on a methodology requires knowledge and judgement about the nature of the impact and risk of interest. Multivariate methodologies, which are not yet mature, need further development for more hazard cases, including treatment of large-scale biases and local statistical characteristics. As these decisions vary on a case-by-case basis, community calibration toolkits would be a valuable resource, for both efficiency of research and fidelity of outputs.

4.2 *Exposure and Vulnerability*

A common challenge in the assessment of climate-related risk is the dynamic nature of exposure and vulnerability, either through shifts in policy, explicit adaptation or both. Building on the UK-SSPs through future work, to provide a broader range of indicators, would allow a more informed assessment of future risk.

Additionally, access to exposure and vulnerability data needs to be improved, as it remains a common and substantive challenge. Information is often sensitive with limited accessibility unless direct partnership with data owners exists. Government data sources are useful assets for informing climate risk assessments but often they are in diverse repositories, with varying formats and access requirements. A broader range of historical and projected future data, curated through an authoritative organisation, on an openly accessible platform where climate data could be either hosted or imported would be a valuable community resource.

Finally, nurturing a community of users that understand their vulnerabilities would be beneficial. Getting credible vulnerability information is regularly the hardest component of the data sourcing for risk, as well as identifying exactly what risk metric(s) are most useful for decision-making. Generally, most organisations have yet to develop the maturity in their understanding and data collection to be able to quantify their vulnerabilities, hindering risk calculations.

5 Conclusions

UKCR has made substantial progress in projections of future exposure and vulnerabilities, and the development of and application of methodologies to combine these with climate projections to quantify future climate risk. Valuable case studies have been produced on agriculture, flooding and overheating, amongst others. However, understanding and quantifying stakeholder vulnerabilities remains a challenge, and access to information needed to estimate exposure and vulnerabilities remains highly fragmented.

The programme has reinforced that different risk frameworks and tools are appropriate for informing different climate resilience and adaptation decisions, and that close communication between stakeholders and climate scientists is crucial to producing valuable analysis and advice.

REFERENCES

1. IPCC. 2021. Climate Change 2021: The Physical Science Basis. Contribution of Working Group I to the Sixth Assessment Report of the Intergovernmental Panel on Climate Change. [Online]. Available at: Climate Change 2021: The Physical Science Basis | Climate Change 2021: The Physical Science Basis (ipcc.ch)).
2. Kendon, M., McCarthy, M., Jevrejeva, S., Matthews, A., Sparks, T., Garforth, J., and Kennedy, J. 2022. State of the UK Climate 2021. *International Journal of Climatology* **42**, pp. 1–80.
3. Climate Change Committee. 2022. Independent Assessment of UK Climate Risk – Advice to Government for the UK's third Climate Change Risk Assessment (CCRA3). [Online]. Available at: https://www.theccc.org.uk/publication/independent-assessment-of-uk-climate-risk/.
4. Simpson, N.P., Mach, K.J., Constable, A., Hess, J., Hogarth, R., Howden, M., Lawrence, J., Lempert, R.J., Muccione, V., Mackey, B. and New, M.G. 2021. A framework for complex climate change risk assessment. *One Earth* **4**(4), pp. 489–501.
5. Glasser, R. 2019. The compounding and cascading consequences of hazards. In: *Preparing for the Era of Disasters* (ed. Glasser, R.) Australian Strategic Policy Institute: JSTOR, pp. 9–11.
6. O'Hare, P. 2021. Manchester's climate risk: a framework for understanding hazards & vulnerability. [Online]. Available at: Manchester Climate Risk: A Framework For Understanding Hazards & Vulnerability | Manchester Climate Change.
7. Arnell, N.W., Freeman, A., Kay, A.L., Rudd, A.C. and Lowe, J.A. 2021. Indicators of climate risk in the UK at different levels of warming. *Environmental Research Communications* **3**(9), p. 095005.
8. Hanlon, H.M., Bernie, D., Carigi, G. and Lowe, J. 2021. Future changes to high impact weather in the UK. *Climatic Change* **166**(3), pp. 1-23.
9. Garry, F.K., Bernie, D.J., Davie, J.C.S andPope, E. 2021. Future climate risk to UK agriculture from compound events. *Climate Risk Management* **32**, 100282.
10. Gupta, R., Howard, A., Davies, M., Mavrogianni, A., Tsoulou, I., Jain, N., Oikonomou, E. and Wilkinson, P. 2021. Monitoring and modelling the risk of summertime overheating and passive solutions to avoid active cooling in London care homes. *Energy and Buildings* **252**, 111418.

11. Kennedy-Asser, A.T., Owen, G., Griffith, G.J., Andrews, O., Lo, Y.E., Mitchell, D.M., Jenkins, K. and Warren, R.F. 2022. Projected risks associated with heat stress in the UK Climate Projections (UKCP18). *Environmental Research Letters* **17**, 034024.
12. Sayers, P., Griffin, A., Lowe, J., Bernie, D., Carr, S., Kay, A. and Stewart, L. 2023. Beyond the Climate Uplift – The importance of accounting for changes in the spatial structure of future fluvial flood events on flood risk in Great Britain. Preprint.
13. Griffin A., Kay, A., Bell, V., Stewart, E., Sayers, P., and Carr, S. 2022. Widespread flooding dynamics changing under climate change: characterising floods using UKCP18. *Hydrology and Earth System Sciences Discussions* **under review**.
14. Griffin, A., Kay A., Stewart, E. and Sayers, P. 2022. Spatially coherent statistical simulation of widespread flooding events under climate change. *Hydrology Research* **53**(11), pp. 1428–1440.
15. Jenkins, K., Kennedy-Asser, A., Andrews, O. and Lo, Y.E. 2022. Updated projections of UK heat-related mortality using policy-relevant global warming levels and socio-economic scenarios. *Environmental Research Letters* **17**, 114036.
16. Garry, F. and Bernie, D. 2023. Characterising temperature and precipitation multi-variate biases in 12 km and 2.2 km UK Climate Projections. *International Journal of Climatology* **43**(6), pp. 2879–2895.
17. Hawkins, E. and Sutton, R. 2009. The potential to narrow uncertainty in regional climate predictions. *Bulletin of the American Meteorological Society* **90**(8), pp. 1095–1108.
18. Jacob, D., Petersen, J., Eggert, B., Alias, A., Christensen, O.B., Bouwer, L.M., Braun, A., Colette, A., Déqué, M., Georgievski, G. and Georgopoulou, E. 2014. EURO-CORDEX: new high-resolution climate change projections for European impact research. *Regional Environmental Change* **14**(2), 563–578.
19. Perks, R. J., Bernie, D., Lowe, J. and Neal, R. 2023. The influence of future weather pattern changes and projected sea-level rise on coastal flooding impacts around the UK. *Climatic Change* **176**(25).
20. Dawkins, L. C., Bernie, D. J., Lowe, J. A. and Economou, T. 2023. Assessing climate risk using ensembles: A novel framework for applying and extending open-source climate risk assessment platforms. *Climate Risk Management* **40**: 100510.

Open Access This chapter is licensed under the terms of the Creative Commons Attribution 4.0 International License (http://creativecommons.org/licenses/by/4.0/), which permits use, sharing, adaptation, distribution and reproduction in any medium or format, as long as you give appropriate credit to the original author(s) and the source, provide a link to the Creative Commons license and indicate if changes were made.

The images or other third party material in this chapter are included in the chapter's Creative Commons license, unless indicated otherwise in a credit line to the material. If material is not included in the chapter's Creative Commons license and your intended use is not permitted by statutory regulation or exceeds the permitted use, you will need to obtain permission directly from the copyright holder.

CHAPTER 12

Note on Delivering Impact

Rachel Harcourt and Nick Hopkins-Bond

Abstract

- Building a strong connection with the target audience, by considering their concerns, priorities, experience and knowledge, is crucial for effective engagement of non-academics in climate change adaptation research.
- Also important is tailoring the method of engagement to each audience and intended purpose; for example, creating visual representations of complex scientific data or undertaking co-creative art projects.

Lead Authors: Rachel Harcourt & Nick Hopkins-Bond

R. Harcourt (✉)
University of Leeds, Leeds, UK
e-mail: r.s.harcourt@leeds.ac.uk

N. Hopkins-Bond (✉)
Met Office, Exeter, UK

© The Author(s) 2024
S. Dessai et al. (eds.), *Quantifying Climate Risk and Building Resilience in the UK*,
https://doi.org/10.1007/978-3-031-39729-5_12

- Considering appropriate and effective ways of measuring impact and benefit from the outset—perhaps co-developed with users—is key to quantifying the overall success of a climate service.

Keywords Engagement · Dissemination · Dialogue · Relationship building · Visuals · Creativity

1 Introduction

The UK Climate Resilience Programme (UKCR) aimed to deliver impact by responding to government priorities and opportunities, and by developing useful and usable research for end users. This book provides evidence of the wealth of outputs produced by the programme. Many UKCR projects also sought to achieve further impact by engaging directly with target audiences. The key learnings from the programme as to how best to do this are summarised in the infographic (Fig. 1) and explored further in the text below. The examples discussed are intended to be illustrative, rather than conclusive, to stimulate ideas and provide guidance for those planning research impact.

2 Ensure Regular Dialogue with End Users Throughout a Project to Ensure All Outputs Are Relevant and Usable

To achieve the key programme aim of developing policy relevant and usable outputs, researchers engaged in dialogue with central and local government partners to understand what 'usable' and 'relevant' means to them. For example, discussions with Bristol City Council as part of the 'Meeting Urban User Needs' project highlighted the need to make UK Climate Projections (UKCP18) data more easily digestible for a non-specialist. The council's requirements were two-fold: to increase the use of UKCP18 in city-level adaptation planning and to create an output to help build risk awareness within the wider Bristol community. Establishing a key point of contact within the research team at an initial face-to-face meeting helped, as city stakeholders were hearing the 'same voice' throughout which nurtured trust. Regular light-touch

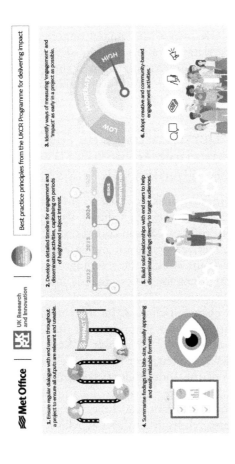

Fig. 1 Best practice principles from the UKCR programme for delivering impact

follow-on meetings with users, interspersed with more in-depth conversations when required, created a continuous knowledge exchange. Through this process of co-development, the partnership produced the Bristol City Pack, an infographic style fact sheet combining simple messaging with an attractive style [1]. Based on feedback from Bristol City Council, this proved extremely successful in providing the required level of accessible and localised climate data. The 'City Pack' format has since been made available to 28 other UK cities with demand ongoing, plus requests for more detailed versions relating to a specific risk, such as heat related health risks in Manchester. This work has highlighted the importance of open dialogue to manage user expectations and communicate what is achievable with the resource and time available.

3 Develop a Detailed Timeline for Engagement and Dissemination Activities Capitalising on Periods of Heightened Subject Interest

When engaging government, it is essential to make use of moments of heightened policy focus. The five-year Climate Change Risk Assessment (CCRA), which is one of the central commitments of the UK 2008 Climate Change Act, receives significant media coverage upon publication. As well as detailing the latest understanding of UK climate risk, the CCRA sets out prioritisations for policy, directly informing the five-year National Adaptation Programme on how the government and others will prepare the UK for a changing climate. Its third iteration, CCRA3, was published in 2021 during the funding period of the UKCR programme, offering a significant opportunity for impact. UKCR researchers undertook a synthesis of UKCR-funded, peer-reviewed research to feed into the Technical Report [2] of the Independent Assessment Evidence Report for the CCRA3 [3]. Due to the protracted nature of the peer review process, it is noted that significant lead times are required to avoid the omission of relevant research from synthesis activities—particularly where it provides evidence for government policy. Factoring this into project timelines is key, and early conversations with publishers could result in special arrangements with regard to publication timescales.

4 IDENTIFY WAYS OF MEASURING 'ENGAGEMENT' AND 'IMPACT' AS EARLY IN A PROJECT AS POSSIBLE

Both 'engagement' and 'impact' can be hard to define, quantify or provide evidence of, particularly during time limited research projects. Some UKCR projects addressed this by issuing feedback surveys and measuring website analytics, although these tend to measure participation rather than impact. Building solid relationships and maintaining dialogue throughout the project, as recommended elsewhere in this note, can facilitate the process of ongoing evaluation and learning. It is also possible to survey the collective impact of diffuse engagement efforts. For example, the UKCR-funded 'RESIL-RISK' project was a national survey of public perceptions of climate risks and adaptation. It found that UK residents are now much more aware of and concerned about the impacts of climate change and extreme weather events, particularly heat stress, in the UK compared with surveys taken only a few years ago. However, the survey did not investigate the likely multiple factors influencing this shift in public perception. Researchers, alongside project partners and intended audiences where possible, should agree success factors and measures of engagement and impact as the project plan is being developed. Although these may change as the project progresses, it is important that measuring impact should not be left to the end.

5 SUMMARISE FINDINGS INTO BITE-SIZE, VISUALLY APPEALING AND EASILY RELATABLE FORMATS

Usability of outputs was a key principle for the UKCR programme. Many policymakers and sector practitioners have no formal climate science training and often have little time to ingest complex climate change information; presentation is therefore vitally important and can be a barrier to impact, even if the information provided is relevant and needed. A UKCR project called 'Communication of Uncertainty' exploring non-experts' responses to written and visual climate information, found that some factors facilitated understanding (e.g. use of colour, simple captions) while others hindered understanding (e.g. use of complex terminology and statistics) [4]. The research team used these findings to develop a set of 'best practice' design principles for communicating climate change information, which they summarised in best practice compliant infographics [4, 5]. The Met Office Communications and Knowledge Integration teams have adopted the guidance as standard practice when designing climate change communication materials, both for UKCR and more widely. The principles provide simple but useful guidance to climate science communicators using all mediums.

6 Build Solid Relationships with End Users to Help Disseminate Findings Directly to Target Audiences

There is a growing demand from many industries for information on current and future climate-related risks, effective adaptation strategies and climate services, driven by the reporting requirements of the UK Climate Change Act 2008, the UK government roadmap for mandatory adoption of the Taskforce on Climate-related Financial Disclosures (TCFD) [6], and ISO standards on adaptation to climate change [7]. One area of focus for UKCR was the UK agricultural sector, which is also responding to new legislation resulting from Brexit, changing expectations regarding land management and challenges due to climate change. This provided real opportunity for impact if UKCR researchers could develop and maintain relationships with those needing information to address this collection of issues. The 'Multiple Hazards' project looked at the risk of compound events (the combined effect of multiple hazards such as temperature and humidity) on the farming sector. From the outset, the research team engaged with government and farming agencies to understand key concerns and priorities, as well as to develop the outputs and types of communications needed. With parallels in many other industries, the relationships were built on the understanding that the agricultural partners brought unique knowledge of their land and experience of farming under variable weather. Building such relationships 'opened doors' to forums and spaces that resonated with the partner industry. In another example, the 'CREWS-UK' project developed a partnership with WineGB, the national association for the English and Welsh wine industry, which provided direct links and influence within the UK wine sector and ensured outputs were relevant to emerging sector priorities.

7 Adopt Creative and Community-Based Engagement Activities

Members of the public have essential roles to play in achieving increased national and local resilience by undertaking adaptive actions in their homes and daily routines, and by supporting adaptation initiatives from government and the private sector. However, climate resilience is a complex issue to communicate, and engaging people on what it means takes an emotional toll. While we are using 'members of the public' here for brevity, work by Climate Outreach [8] and others has shown that different sections of the UK public have very differing needs, interests and preferences. One means of addressing this is to bring creative climate communications into spaces where people already are. For example, the 'Risky Cities' project based in Hull drew the attention of city residents to local flood risks by exhibiting large-scale light and sound installations. In doing so, the project connected to its intended audience through the shared language of 'Hull' and used art to explore the city's long history of living with water. In other examples, 'Creative Climate Resilience' worked with Manchester neighbourhoods to co-develop creative outputs that explored the temporal and geographical relevance of climate resilience to local communities, and another project, 'Time and Tide' conducted interactive performances on beaches and exhibitions in coastal communities. Researchers noted that these approaches can develop a sense of agency and ownership in the affected communities, while also bringing joyfulness and playfulness into a conversation which is often emotionally demanding. Chapters 3 and 6 provide further information on this topic.

Acknowledgements The authors would like to thank the following UKCR researchers for sharing their insights into delivering impact through their projects: Jenna Ashton, Richard Betts, Ed Brookes, Kate Gannon, Freya Garry, Helen Hanlon, Paul O'Hare, Claire Scannell and Corinna Wagner. Thanks also to George Burningham and Phoebe Wu for help with producing the infographic.

References

1. Met Office. 2021. Bristol 'City Pack'. [Online]. Available at: SPF City Pack_editable_template (ukclimateresilience.org).
2. Betts, R.A., Haward, A.B. and Pearson, K.V. (eds) 2021. *The Third UK Climate Change Risk Assessment Technical Report*. [Online]. Available at:

Technical-Report-The-Third-Climate-Change-Risk-Assessment.pdf (ukclimaterisk.org).
3. Climate Change Committee. 2022. Independent Assessment of UK Climate Risk – Advice to Government for the UK's third Climate Change Risk Assessment (CCRA3). [Online]. Available from: https://www.theccc.org.uk/publication/independent-assessment-of-uk-climate-risk/.
4. Kause, A., Bruine de Bruin, W., Fung, F., Taylor, A. and Lowe, J. 2020. Visualisations of projected rainfall change in the United Kingdom: An interview study about user perceptions. *Sustainability* **12**(7), 2955.
5. Kause, A., Bruine de Bruin, W., Domingos, S., Mittal, N., Lowe, J. and Fung, F. 2021. Communications about uncertainty in scientific climate-related findings: a qualitative systemic review. *Environmental Research Letters* **16**(5), 053005.
6. HM Treasury. 2020. A Roadmap towards mandatory climate-related disclosures. [Online]. Available at: https://assets.publishing.service.gov.uk/government/uploads/system/uploads/attachment_data/file/933783/FINAL_TCFD_ROADMAP.pdf.
7. International Organization for Standardization. 2021. ISO 14091:2021 Adaptation to climate change: Guidelines on vulnerability, impacts and risk assessment, Geneva: ISO.
8. Climate Outreach. The Seven Segments in Depth [Online]. Available from: https://climateoutreach.org/britain-talks-climate/seven-segments/.

Open Access This chapter is licensed under the terms of the Creative Commons Attribution 4.0 International License (http://creativecommons.org/licenses/by/4.0/), which permits use, sharing, adaptation, distribution and reproduction in any medium or format, as long as you give appropriate credit to the original author(s) and the source, provide a link to the Creative Commons license and indicate if changes were made.

The images or other third party material in this chapter are included in the chapter's Creative Commons license, unless indicated otherwise in a credit line to the material. If material is not included in the chapter's Creative Commons license and your intended use is not permitted by statutory regulation or exceeds the permitted use, you will need to obtain permission directly from the copyright holder.

CHAPTER 13

Afterword

Suraje Dessai, Kate Lonsdale, Jason Lowe and Rachel Harcourt

Abstract

- Investing in 'gluing' roles, as was performed by the UKCR Champions, is essential for building community and delivering impact, as is a strong online and social media presence and a programme of community building events.

S. Dessai (✉) · K. Lonsdale · R. Harcourt
University of Leeds, Leeds, UK
e-mail: s.dessai@leeds.ac.uk

K. Lonsdale
e-mail: kate.lonsdale@climatesense.global

R. Harcourt
e-mail: r.s.harcourt@leeds.ac.uk

J. Lowe
Met Office, Exeter, UK
e-mail: jason.lowe@metoffice.gov.uk

© The Author(s) 2024
S. Dessai et al. (eds.), *Quantifying Climate Risk and Building Resilience in the UK*,
https://doi.org/10.1007/978-3-031-39729-5_13

- Resilience research should be 'user' or 'challenge' led and needs to invest in ways of working that facilitate innovative and transdisciplinary approaches. Key research gaps remain including in understanding compound, transboundary, cascading and systemic risks; place-based vulnerability assessments that combine risk information with other socioeconomic and behavioural factors; and scaling-up climate services.
- Future adaptation research programmes should prioritise further developing the research-practice community to adequately address the complex challenge of building resilience.

Keywords community building · delivering impact · transdisciplinary research · research gaps

Between 2019 and 2023, we championed the Strategic Priorities Fund (SPF) UK Climate Resilience Programme (UKCR), including the production of this volume. This gave us a unique perspective on the research, practice and policy of climate resilience in the UK. Here, we reflect on this experience and present some key messages. In Sect. 1, we reflect on ways of working and community building. In Sect. 2, we summarise the programme's achievements in producing novel evidence for climate resilience. In Sect. 3, we reflect on lessons learned about interdisciplinary and transdisciplinary approaches to climate resilience research. In Sect. 4, we focus on what remains to be done to address ongoing research questions, and also how to design and deliver fit for purpose research to enhance resilience building.

1   Ways of Working and Community Building

The UKCR Programme and Science Plan recognised the importance of stakeholders and end users in the development of useful and usable climate resilience research. One of the programme's aims was to grow the community of interacting researchers, practitioners and policymakers in climate resilience. As a result, the funded work of UKCR included a wide spectrum of approaches to co-development and co-production (see Chapter 3).

The role of the Champions, supported by the wider Champion team and the Met Office science coordinator, has been critical in building and maintaining this community. Many aspects of all these roles have been about building relationships with, and making connections between, funded projects and with target groups and initiatives. The importance of investing in people to play such 'gluing' roles should not be underestimated; this investment has acted to maximise the value and impact of interdisciplinary and transdisciplinary research programmes, and mirrors messages emerging from UKCR projects about the need to build trusting relationships at every level. This is essential if we are to adequately address the complex issue of building climate resilience. As the programme comes to an end, the Champion team's role has increasingly focused on the synthesis and tailoring of key messages for specific audiences, in order to maximise the impact of the programme for ongoing policy and practice and to ensure its longer term legacy.

Community building started well with a programme-wide workshop in November 2019 held in Leeds, which convened the first tranche of funded projects from both the Met Office and UK Research and Innovation (UKRI). Here, participants identified cross-cutting themes for the programme. Only a few months later, the 'work from home' requirement of the COVID-19 pandemic limited the extent to which the UKCR community could interact 'in person'. Activities to enhance connections across and beyond the funded work of the programme had to be rapidly rethought and moved online. For example:

- **Fortnightly webinars** were established, whereby academic research teams could share initial findings. The format of the webinars then allowed for a response from a non-academic partner or beneficiary who could give their perspective on the usefulness and relevance of the research, followed by a Q&A session.[1]
- **Quarterly virtual forums** were arranged, to share project updates and relevant news, and discuss specific topics.
- **A mid-term, two-day online conference** was organised with the Climate Change Committee and National Centre for Atmospheric Science, to examine if the UK is on track to adapt to climate

---

[1] https://www.youtube.com/playlist?list=PLgyCRS_bWUxoJuZ5MueERVDf62S76ZnuJ

change. Over 300 invited participants debated the climate science and possible climate impacts, how far current and planned adaptation efforts go to manage the risks, and what more would need to be done to prepare (https://www.ukclimaterisk.org/learn-more/conference-is-the-uk-on-track-to-adapt-to-climate-change/).
- **An online Programme Assembly** was organised, to help guide the direction of the programme, such as the priorities for synthesis (https://www.ukclimateresilience.org/wp-content/uploads/2021/10/UKCR-Assembly-Sept-21-Workshop-Report-FINAL_6p.pdf).

The effectiveness of these activities was greatly enhanced by the skilled and experienced communications support in the Champion team and a dedicated programme website. This not only provided news, blogs and information about the funded work and a popular archive of webinars, but also community links through social media (the programme has over 2,200 followers on Twitter/X) and a regular newsletter that reached over 2,250 subscribers.

As the pandemic eased and UKCR entered its final year, programme activities focused on programme-level synthesis of messages on common themes from the funded work, including two end of programme events. The first was an in-person 'Showcase' (https://www.ukclimateresilience.org/news-events/climate-adaptation-project-outputs-showcased-in-hull/) in Hull, in October 2022, designed to celebrate the work of the programme through performances, tool demonstrations and opportunities to discuss enhanced application of programme findings and outputs and stimulate discussion on climate risks and how to manage them. The second event, an end of programme conference (https://www.ukclimateresilience.org/ukcr-final-conference/) in London in March 2023, presented the programme's research advancements and discussed its implications for policy and practice.

Many UKCR projects were both interdisciplinary (involving several academic disciplines) and transdisciplinary (involving stakeholders in knowledge production). The academic disciplines involved included the arts and humanities, engineering, social science and natural science. The nature of some UKCR projects required cross-community participation to develop their outputs, such as the National Framework for Climate Services (https://www.ukclimateresilience.org/wp-content/uploads/2022/11/Recommendations-UK-NFCS-AUG22.pdf) and a guidance standard for climate services.[2] This cross-community collaboration also helped to build connections and share experience. Another collaborative, community-wide task was the programme's contribution to the Technical Report of the third Climate Change Risk Assessment (CCRA3); UKCR co-funded the project lead and developed an open access special issue of the journal *Climate Risk Management* [1] (https://www.sciencedirect.com/journal/climate-risk-management/special-issue/105D9F0R4PQ), a compilation of new research on UK climate risk assessment and management to support the evidence base for CCRA3.[3]

2   Novel Evidence

The SPF UKCR programme produced a range of novel research outputs across the three research themes: characterising and quantifying climate-related risks, managing climate-related risks through adaptation and co-producing climate services. These outputs are already enhancing the UK's capacity for climate risk assessment and improving the nation's climate resilience. The range of outputs are described earlier in this volume; here, we highlight areas of particular novelty and progress that form an important part of the UKCR programme's legacy.

An important aspect of UKCR has been the inter- and transdisciplinary nature of many of the projects that is essential for addressing real-world problems that inevitably cut across academic disciplinary boundaries. This has contributed to our understanding of how people, organisations and policy contribute to adaptation at different scales. For example, projects have provided new understanding of community-based flood resilience [2] (https://www.communityactionforwater.org/) and adaptive responses for both staff and residents in care settings [3]. Several projects (https://riskycities.hull.ac.uk/) and embedded researchers trialled innovative arts and humanities approaches to build climate awareness and agency, helping communities reflect on identity, loss and learning from the past in order to become more flood

---

[2] https://www.ukclimateresilience.org/wp-content/uploads/2021/01/Climate-Services-Standard-Final-for-Publication.pdf

[3] Of the 12 published papers, eight were funded by UKCR.

resilient. New learning on approaches to co-producing knowledge and communicating risks was also developed, as detailed in Chapters 3–6.

The programme has advanced climate service development and delivery in the UK through establishing a roadmap for a national climate service (https://www.ukclimateresilience.org/wp-content/uploads/2022/11/Recommendations-UK-NFCS-AUG22.pdf), aligned with the Global Framework for Climate Services. Other achievements include developing demonstrator climate services and decision support tools, for example, an urban service (https://www.ukclimateresilience.org/projects/prototype-development-meeting-urban-user-needs/) that has delivered climate city packs to 30 UK councils to raise awareness of and manage climate risks. Support for future climate services has been enhanced through a new, fully tested toolkit, which will help scale up pilot projects to reach a wider range of users, plus a voluntary standard to improve the quality of climate services (https://www.ukclimateresilience.org/projects/climate-services-standards-and-value/). For more on climate services and decision support tools, see Chapters 7 and 8.

Novel aspects of the research relating to an improved understanding of climate hazards and risks include demonstrating the application of event attribution to more impact relevant metrics. For instance, for extreme rainfall [4] and heat-related mortality [5] UKCR has improved the characterisation of compound hazards, including joint surge and river flooding around the UK coastline [6, 7] and for agricultural relevant impacts [8]. An important new dataset provided in UK Climate Projections (UKCP18) comes from the climate simulations made using convective permitting models, and UKCR has been able to exploit this new data to better understand urban interactions with climate [9] and the future evolution of intense storms. Further novel hazard-related work in UKCR has provided a toolkit for estimating sea-level rise along the UK coastline, which is relocatable around the globe, and a new dataset of river flows and drought metrics for the UK [10]. Finally, UKCR has also produced a new resource of analysed EURO-CORDEX climate hazard data for the UK region allowing it to be used alongside UKCP18 climate results to better sample uncertainty in future projections. For further information, see Chapter 9 of this volume.

Another area of innovation, and a missing ingredient from previous UK risk assessment, is a national scale set of socioeconomic scenarios that are consistent with global and regional shared socioeconomic pathways that are used in many international climate studies [11–13]. These scenarios are enabling improved treatment of future exposure and vulnerability in UK risk assessments, as referenced in Chapters 3, 7 and 11.

Alongside the advances in hazard and socioeconomic scenarios, UKCR has produced a step change in climate change risk assessment capability with the development of a new framework (OpenCLIM). This enables a linked set of existing and new spatially explicit impact models to be driven by consistent sets of climate and socioeconomic data and adaptation, interventions. It can be applied at a range of scales from national to local and will facilitate more quantitative spatial modelling in future national risk assessments. Other key developments in spatial risk assessment by UKCR include the use of catastrophe modelling techniques, more usually applied in the insurance industry, to issues of longer term climate risk and resilience decision-making. These new methods complement datasets of risk informed hazard metrics for a range of emission scenarios and global warming levels that were produced in the earlier stages of UKCR and used widely in CCRA3, and by a range of organisations via web interfaces. See Chapters 10 and 11 for more.

3 Reflections on Developing a Transdisciplinary Research Programme

All SPF-funded research had a requirement to link to government research and innovation priorities, which for UKCR included the Department for Environment, Food and Rural Affairs (Defra) and the Climate Change Committee (CCC). The programme used the Fig. 1 to think through how to enhance research usefulness and usability. The diagram illustrates the range of approaches that can be taken by research projects and programmes to encourage the use of research outputs. 'Linear dissemination' and the left end of the spectrum can be achieved through 'knowledge products' (e.g. academic papers, briefing notes and tailored information packs). As you move towards the 'co-production' end of the spectrum, the increasing importance of human and relational skills in knowledge brokering come to the fore. Here, the roles become more about convening conversations, building relationships and sharing practice-based (and more 'tacit') knowledge that may be harder to capture and share through knowledge products. This again emphasises the importance of the 'glue' roles and mechanisms needed to convene, signpost, connect and synthesise across different projects, systems and organisations, to ensure the programme was more than simply the sum of its parts (Fig. 1).

The UKCR Science Plan, and the legacy items described within it, provided an overview of what the programme aimed to achieve. Given the focus of SPF funding, annual and final programme evaluations augmented the criterion for 'research excellence', demonstrated by the production and citation of peer-reviewed papers, with additional criteria such as effectiveness of partnerships and co-production and the value of the research to intended users. The Champion team also managed a small, flexible fund for synthesis activities that could be drawn on to enhance the accessibility of project findings, or to synthesise findings of different projects on a common theme.[4] The programme, therefore, awarded small amounts of additional funding for synthesis and engagement (e.g. through infographics), recognising the importance of targeting user groups through non-academic means.

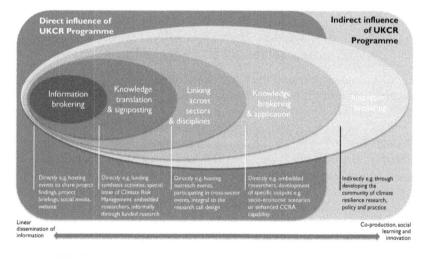

Fig. 1 UKCR's knowledge brokering, translation and application roles building on [14–16]

---

[4] Example, https://www.youtube.com/watch?v=UgSvmxczbgc&t=18s.

Several projects attempted some degree of intentional co-production (see Chapter **3**) to ensure research outputs are usable by policymakers and practitioners. As discussed in that chapter, 'co-production' captures a broad spectrum of approaches, each with challenges limiting the extent to which long-term relationships can be established and maintained, although much of this is solvable through new approaches to research.

An expectation for non-academic partner involvement in the research was set up through the requirements of the research calls, e.g. stating the research must have 'strong stakeholder engagement through the research process' or 'clear co-design, co-production and vision for creating impact'. Careful consideration of research design to enhance mutual learning was encouraged from early in the research process when there was the greatest opportunity for shaping goals, outputs and approach. Projects found that the research scoping phase was also an opportunity to share understanding of core concepts and language, and agree ways of working to ensure alignment across differing organisational incentive structures and cultures. Periodic 'pause points' to jointly review progress provided additional opportunities to check the research remained fit for purpose for all.

The Embedded Researcher scheme (see Chapter **4**) was deliberately designed to enhance the relevance and use of outputs, shifting the traditional approach of academic-led identification of research questions to enable non-academic partners to state their research needs and ensure the research was designed to meet them. Host organisations, including city councils, government departments, agencies, non-governmental organisations and the Church of England, enriched the UKCR network, both contributing to and benefiting from access to the wider programme.

The new connections across disciplines and with non-academic partners needed for transdisciplinary research take time to build. As funding bid schedules were often rushed to fit funding deadlines, this limited the creative potential of new relationships and the design of innovative responses to research calls. Non-academic partners were included in programme activities where possible, for example, chairing funding panels and responding to academic webinars. These inputs provided an important 'reality check' on the significance of the funded research for practice.

Standard UKRI funding is only available for researchers, and therefore, non-academic partners had to be self-funding, creating an imbalance of influence on the research focus and design. While the Met Office was less restricted in bringing in non-academics through external calls, it was not possible to have common funding pots between the two organisations. This limitation was navigated by allocating different funding types to UKRI and Met Office as appropriate, which included open research calls, single tenders and open tenders. Like the Embedded Researcher scheme, open tenders funded through the Met Office enabled non-academics, such as consultancies, to bid for the work. Future programmes would benefit from providing a more even playing field for non-academic and academic partners.

4 Research Gaps and Future Directions

While the SPF UKCR programme has made significant steps forward in the consideration of risk and resilience in the UK, there remains much to do. Given limited time and budget, the programme had to make choices about where to focus to have most significant and lasting impact. Here, we offer thoughts on future priorities, in terms of both *what* to research and *how* to do it. We hope this will be useful as we enter the fourth UK Climate Change Risk Assessment and National Adaptation Programme cycle.

4.1 Transdisciplinary Research

There is a need to transform resilience research from being fragmented and siloed, to collaborative, learning oriented, just, inclusive, embracing of diverse sources of knowledge, contextualised and experimental [17]. The term 'transdisciplinarity' is an academic construct, not recognised by the organisations and networks having to respond to our changing climate. They start with a problem or challenge and build from there. Shifting academic research to a 'user' or 'challenge' focus requires funding and research models that enable much greater collaboration between all disciplines and enable private and public sector colleagues, policymakers and the public to participate on an equal footing with researchers. For example, funding an initial scoping stage to convene interested parties could support imaginative, co-created and transdisciplinary responses to research calls.

4.2 Boundary-Spanning Skills

One of the most pressing needs is the translation of climate science into information that is useful and usable for those tasked to make the UK more climate resilient. This will require a significant scaling-up of boundary-spanning skills, including undertaking co-production, working in transdisciplinary teams, scoping and defining the problem and translating science for users; skills which are currently often treated as of secondary importance after subject knowledge. This should start in undergraduate teaching and through the continuous professional development of researchers, practitioners and those in related industries. Further to the general upskilling of the climate resilience community, there will be increased need for 'science translators', likely specialising in different audiences—for instance, what national policymakers need from the science community may differ from what local government needs. These roles need to be budgeted for in research departments and funding bids.

4.3 Managing the Risk

Understanding how risks can be reduced through resilience building and adaptation is a priority. While the physical science aspect of this has made strides in the programme, the next phase needs better integration of socioeconomic and behavioural factors, including social inequalities and vulnerability. Location-specific research is still critical for effective adaptation, particularly how to meaningfully include the affected communities from the project planning stage onwards, as well as work on the transferability of adaptation lessons across locations. Capturing the case experience of barriers to and enablers of good adaptation practice at different scales (regional, sectoral and organisational) helps to shed light on why, despite greater understanding and awareness of climate risk, it is still challenging to translate this into adaptation strategy and operational plans. More focus on leadership and governance as a driver for greater action, and the integration of mitigation and adaptation, is now needed.

4.4 Co-producing Climate Services

While the programme has demonstrated prototype climate services and lessons on scaling services [18], this must now be put into practice through, for example, innovation accelerators. There is still a tendency for a 'science-first' approach, and while co-development of risk and resilience projects is more widespread, more could be done to ensure that the work is embedded in relevant decision contexts. As implementation proceeds, more work will be needed to monitor, evaluate and learn from resilience building initiatives to both track progress and gather good practice. Further support for the standard for climate services (e.g. converting it to BSI or ISO standards) and implementing the proposed national framework for climate services, aligned with the World Meteorological Organization's global framework, would add great value to the coherence and quality of the UK's climate services sector.

4.5 Hazard to Risk

The methodologies and datasets developed under UKCR need to be integrated into new risk estimates for the entire UK. This includes improved understanding of storm characteristics from higher resolution models and approaches for dealing with compound impacts on the land and at the coast. There remain significant gaps around transboundary risks, systemic risks and cascading impacts. In particular, there is still a lack of diversity of research approaches to quantifying the system risks that follow from direct climate impacts onto the UK. Improving this could be usefully applied at a spatial detail relevant to adaptation while covering the entire geography of the UK and all sectors of activity. There also remains a need to understand the consistency of the different approaches and datasets. One key UKCR project, 'OpenCLIM', has produced a framework that is helping to establish a more consistent approach to place-based risk assessment, but this should now be expanded to a greater range of risks and bring in a wider range of component models from other risk and resilience research.

5 Concluding Thoughts

The UK Climate Resilience programme has improved our collective understanding of the climate risks we face and the implications of those risks, as well as increasing the availability of tools and information needed to assess them. It has also created a community of interacting researchers, practitioners and policymakers in climate resilience.

There now needs to be a shift in focus towards strengthening the UK's capacity to adapt. This requires a significantly larger effort and a more diverse set of actors. We urge future research programmes to appreciate the value of a connected research-practice community for climate resilience, to build strong relationships between academia and organisational or policy practice that allow for quick and clear feedback loops to ensure outputs are relevant and usable. Future programmes should include innovators and entrepreneurs, skilled in developing research insight into practice through innovation hubs and platforms.

The UK Climate Resilience Programme has shown the potential of such a community to advance thinking and practice on climate resilience in the UK. There is considerable appetite to build on this experience, to ensure the UK sustains and enhances this progress for the public good.

REFERENCES

1. Dessai, S., Fowler, H.J., Hall, J.W. and Mitchell, D.M. 2022. UK Climate Risk Assessment and Management. *Climate Risk Management* **37**, 100440.
2. Sefton, C., Sharp, L., Quinn, R., Stovin, V. and Pitcher, L. 2022. The feasibility of domestic raintanks contributing to community-oriented urban flood resilience. *Climate Risk Management* **35**, 100390.
3. Ibbetson, A., Milojevic, A., Mavrogianni, A., Oikonomou, E., Jain, N., Tsoulou, I., Petrou, G., Gupta, R., Davies, M. and Wilkinson, P. 2021. Mortality benefit of building adaptations to protect care home residents against heat risks in the context of uncertainty over loss of life expectancy from heat. *Climate Risk Management* **32**, 100307.
4. Cotterill, D., Stott, P., Christidis, N. and Kendon, E. 2021. Increase in the frequency of extreme daily precipitation in the United Kingdom in autumn. *Weather and Climate Extremes* **33**, 100340.
5. Huang, W.T.K., Braithwaite, I. Charlton-Perez, A., Sarran, C. and Sun, T. 2022. Non-linear response of temperature-related mortality risk to global warming in England and Wales. *Environmental Research Letters* **17**(3), 034017.
6. Robins, P.E., Lewis, M.J., Elnahrawi, M., Lyddon, C., Dickson, N. and Coulthard, T.J. 2021. Compound Flooding: Dependence at Sub-daily Scales Between Extreme Storm Surge and Fluvial Flow. *Frontiers in Built Environment* **7**, 727294.
7. Harrison, L.M., Coulthard, T.J., Robins, P.E. and Lewis, M.J. 2022. Sensitivity of Estuaries to Compound Flooding. *Estuaries and Coasts* **45**(5), pp. 1250–1269.
8. Garry, F.K., Bernie, D.J., Davie, J.C.S. and Pope, E.C.D. 2021. Future climate risk to UK agriculture from compound events. *Climate Risk Management* **32**, 100282.

9. Keat, W.J., Kendon, E.J., and Bohnenstengel, S.I. 2021. Climate change over UK cities: the urban influence on extreme temperatures in the UK climate projections. *Climate Dynamics* **57**(11), pp. 3583–3597.
10. Hannaford, J., Mackay, J., Ascott, M., Bell, V., Chitson, T., Cole, S., Counsell, C., Durant, M., Jackson, C. R., Kay, A., Lane, R., Mansour, M., Moore, R., Parry, S., Rudd, A., Simpson, M., Facer-Childs, K., Turner, S., Wallbank, J., Wells, S., and Wilcox, A. 2023. The enhanced future Flows and Groundwater dataset: Development and evaluation of nationally consistent hydrological projections based on UKCP18. *Earth System Science Data Discussions* **15**(6): 2391–2415.
11. Merkle, M., Alexander, P., Brown, C., Seo, B., Harrison, P.A., Harmáčková, Z.V., Pedde, S. and Rounsevell, M., 2022. Downscaling population and urban land use for socio-economic scenarios in the UK. *Regional Environmental Change* **22**, 106.
12. Pedde, S., Harrison, P., Holman, I., Powney, G.D., Lofts, S., Schmucki, R., Gramberger, M. and Bullock, J.M. 2021. Enriching the Shared Socioeconomic Pathways to co-create consistent multi-sector scenarios for the UK. *Science of the Total Environment* **756**, 143172.
13. Harmáčková, Z.A., Pedde, S., Bullock, J.M., Dellaccio, O., Dicks, J., Linney, G., Merkle, M., Rounsevell, M.D.A., Stenning, J. and Harrison, P.A. 2022. Improving regional applicability of the UK shared socioeconomic Pathways through iterative participatory co-design. *Climate Risk Management* **37**, 100452.
14. Fisher, C. 2011. *Knowledge Brokering and Intermediary concepts. Analysis of an e-discussion on the Knowledge Brokers Forum.* Institute of Development Studies, Knowledge Services, Brighton, UK.
15. Harvey, B., Lewin, T. and Fisher, C. 2012. Introduction: Is Development Research Communication Coming of Age? *IDS Bulletin* **43**(5), pp. 1–8.
16. Shaxson, L., Bielak, A., Ahmed, I., Brien, D., Conant, B., Fisher, C., Gwyn, E., Klerkx, L., Middleton, A., Morton, S., Pant, L. and Phipps, D. 2012. Expanding our understanding of K* (KT, KE, KTT, KMb, KB, KM, etc.). A concept paper emerging from the K* conference held in Hamilton, Ontario, Canada, April 2012. *United Nations University, Institute for Water, Environment and Health* pp. 1–88.
17. Fazey, I. et al. 2020. Transforming knowledge systems for life on Earth: visions of future systems and how to get there. *Energy Research and Social Science* **70**, 101724.
18. Gunetchev, G., Palin, E.J., Lowe, J. and Harrison, M. 2023. Upscaling of climate services - what is it? A literature review. *Climate Services* **30**, 100352.

Open Access This chapter is licensed under the terms of the Creative Commons Attribution 4.0 International License (http://creativecommons.org/licenses/by/4.0/), which permits use, sharing, adaptation, distribution and reproduction in any medium or format, as long as you give appropriate credit to the original author(s) and the source, provide a link to the Creative Commons license and indicate if changes were made.

The images or other third party material in this chapter are included in the chapter's Creative Commons license, unless indicated otherwise in a credit line to the material. If material is not included in the chapter's Creative Commons license and your intended use is not permitted by statutory regulation or exceeds the permitted use, you will need to obtain permission directly from the copyright holder.

Project References

This section provides further detail for each of the UK Climate Resilience projects referred to in this volume.

Listed for each are: project title (followed by extended title where applicable), project aim, principal investigators and collaborating institutes, project webpage(s) and funding tranche.

AquaCAT (Flood risk estimates using techniques from catastrophe modelling)

This project aims to combine information from UKCP18 with catastrophe modelling and apply it to climate-driven changes in UK flood risk.

Paul Sayers (Sayers and Partners), with UK Centre for Ecology and Hydrology, Vivid Economics Loughborough University

https://www.ukclimateresilience.org/projects/flood-risk-estimates-using-techniques-from-catastrophe-modelling/

Met Office—Improving Climate Hazard Information, from Climate Hazard to Climate Risk

ARID (ARID: School buildings adaptation, resilience and impacts on decarbonisation in a changing climate)

This project aims to develop risk-informed resilience of school building stock and optimise the opportunities from a transition to a low carbon future.

Daniel Godoy Shimizu (University College London), with Department for Education https://www.ukclimateresilience.org/projects/arid-school-buildings-adaptation-resilience-and-impacts-on-decarbonisation-in-a-changing-climate/

Embedded Researcher Cohort 1

Bristol Heat Resilience (Developing an urban heat resilience plan for Bristol)

This project aims to co-develop a heat vulnerability index and heat resilience plan for Bristol, to support climate resilience strategies.

Charlotte Brown (University of Manchester), with Bristol City Council

https://www.ukclimateresilience.org/projects/developing-an-urban-heat-resilience-plan-for-bristol/

Embedded Researcher Cohort 1

Catchment Erosion Resilience (Erosion hazards in river catchments: Making critical infrastructure more climate resilient)

This project aims to model how future climate scenarios will affect erosion hazards in river catchments and vulnerability of associated infrastructure.

James Cooper (University of Liverpool), with CoirGreen, ARUP, Waterco

https://www.ukclimateresilience.org/projects/erosion-hazards-in-river-catchments-making-critical-infrastructure-more-climate-resilient/

Phase 1 Projects

CLandage (Building climate resilience through community, landscapes and cultural heritage)

This project aims to use learning from the past to better understand how communities might adapt to future changes in places and landscapes.

Neil McDonald (University of Liverpool), with Historic England, University of Glasgow, Northumbria University, Staffordshire Record Office, Museum & Tasglann nan Eilean Siar, Fjordr

https://www.ukclimateresilience.org/projects/clandage-building-climate-resilience-through-community-landscapes-and-cultural-heritage/

https://historicengland.org.uk/whats-new/research/building-climate-resilience-through-community-landscapes-and-cultural-heritage/

Living with Uncertainty

ClimaCare (Governing the climate adaptation of care settings)

This project aims to quantify climate-related heat risks in care settings nationwide.

Mike Davies (University College London), with London School of Hygiene and Tropical Medicine, Oxford Brookes University, Care Quality

Commission, Ministry of Housing, Communities and Local Government, Chartered Institution of Building Services Engineers, the Greater London Authority, Aston House, PRP, Met Office

https://www.ukclimateresilience.org/projects/climacare-governing-the-climate-adaptation-of-care-settings/

Governing Adaptation

Climate Information for Decision-Making (Climate information to Inform UK decision-making)

This project aims to determine what is needed to support UK climate risk assessment and adaptation decision-making over the next decade, through engagement with users and providers.

Murray Dale (JBA Consulting), with Cardiff University, Becky Venton

https://www.ukclimateresilience.org/projects/climate-information-to-inform-uk-decision-making/

Met Office—Improving Climate Hazard Information, from Climate Hazard to Climate Risk

Climate Resilience Standards (Review of standards, guidance and codes of practice for enhancing climate resilience)

This project aims to understand the climate information inputs into commonly used national guidance standards.

Murray Dale (JBA Consulting Ltd), John Dora Consulting Limited, TRIOSS, University of Leeds, British Standards Institute

https://www.ukclimateresilience.org/projects/review-of-standards-guidance-and-codes-of-practice-for-enhancing-climate-resilience/

Met Office—Pilot Climate Services

Climate Risk Indicators (Developing indicators of climate risk using UKCP18 to support risk assessments and enhance resilience)

This project aims to provide first estimates of a series of indicators of climate risk, relevant to national, regional and local climate risk assessments.

Nigel Arnell (University of Reading), with UK Centre for Ecology and Hydrology, University of Leeds

https://www.ukclimateresilience.org/projects/climate-risk-indicators-developing-indicators-of-climate-risk-using-ukcp18-to-support-risk-assessments-and-enhance-resilience/

https://uk-cri.org/

Phase 1 Projects

Climate Services Standards (Climate services standards and value)

This project aims to establish a coherent set of standards for climate services, so decision-makers can improve their capacity to manage climate-related risk.

Murray Dale (JBA Consulting Ltd), with Climate Sense, Paul Watkiss Associates, Becky Venton, Prof Rob Wilby

https://www.ukclimateresilience.org/projects/climate-services-standards-and-value/

Met Office—Operational Climate Services

Climate Stress Testing (Climate stress testing the UK food supply chain using earth observation)

This project aims to bring together UK food supply chain stakeholders with earth observation researchers, to create climate stress testing tools to improve UK food security.

Caitlin Douglas (King's College London), with Space4Climate, London Climate Change Partnership

https://www.ukclimateresilience.org/projects/climate-stress-testing-the-uk-food-supply-chain-using-earth-observation/

Embedded Researcher Cohort 1

Coastal Climate Services (Climate service pilot: Improving coastal resilience)

This project aims to co-develop with users a coastal resilience service, by further developing the existing UKCP18 Sea Level Rise tool.

Rachel Perks, Dan Bernie (Met Office)

https://www.ukclimateresilience.org/projects/climate-service-pilot-improving-coastal-resilience/

Met Office—Pilot Climate Services

CoastalRes (Coastal resilience in the face of sea level rise: Making the most of natural systems)

This project aims to develop and demonstrate prototype methods to assess realistic pathways for strategic coastal erosion and flood resilience in response to climate change.

Robert Nicholls (University of Southampton) with University College London, Middlesex University, National Flood Forum, National Trust, Network Rail, Wildfowl and Wetlands Trust, ABPmer, Natural England

https://www.ukclimateresilience.org/projects/coastal-resilience-in-the-face-of-sea-level-rise-making-the-most-of-natural-systems/

https://www.coastalmonitoring.org/ccoresources/coastalres/

Phase 1 Projects

Creative Climate Resilience (Community climate resilience through folk pageantry)

This project aims use community knowledge to deliver a case study that responds to Manchester's climate action policies and community contexts.

Jenna Ashton (University of Manchester), with Manchester Climate Change Agency, Manchester City Council, Neighbourhoods North Manchester, Northern Chamber Orchestra, National Trust North Region, Manchester Arts and Sustainability Team, Manchester Institute of Education, Black Environment Network, A Bird in the Hand Theatre, Manchester Environment Education Networkhttps://www.ukclimateresilience.org/projects/community-climate-resilience-through-folk-pageantry/

https://creative-climate-resilience.org/

Living with Uncertainty

CREWS-UK (Characterising and adapting to climate risks in the UK wine sector)

This project aims to generate practical support for climate resilience in the UK, particularly for the wine sector.

Declan Conway (London School of Economics), with University of East Anglia, Wines of Great Britain

https://www.ukclimateresilience.org/projects/crews-uk-characterising-and-adapting-to-climate-risks-in-the-uk-wine-sector-climate-resilience-in-the-uk-wine-sector/

https://www.lse.ac.uk/granthaminstitute/resilient-wine/

Phase 1 Projects

CROP-NET (Monitoring and predicting the effects of climate change on crop yields)

This project aims to scope out the requirements for a robust, real-time crop and grass yield monitoring and modelling service for the UK to provide improved predictions of future climate change impacts.

Richard Pywell (UK Centre for Ecology & Hydrology), with University of Reading, University of Leeds

https://www.ukclimateresilience.org/projects/crop-net-monitoring-and-predicting-the-effects-of-climate-change-on-crop-yields/

https://cropnet-demonstrator.datalabs.ceh.ac.uk/

Phase 1 Projects

eFLaG (Prototype development: enhancing the resilience of the water sector to drought events)

This project aims to co-develop a pilot climate service to ensure a coherent, national approach to ensure drought resilience for the UK's water sector under a changing climate.

Jamie Hannaford (UK Centre for Ecology and Hydrology) and Chris Counsell (HR Wallingford)

https://www.ukclimateresilience.org/projects/enhancing-the-resilience-of-the-water-sector-to-drought-events-climate-service-pilots/

Met Office—Pilot Climate Services

Environment Agency Incident Response (Adapting Environment Agency incident response for climate resilience)

This project aims to characterise and quantify Environment Agency flood and drought incident response activity in current and future climates.

Elizabeth Lewis (Newcastle University), with Environment Agency

https://www.ukclimateresilience.org/projects/adapting-environment-agency-incident-response-for-climate-resilience/

Embedded Researcher Cohort 1

EuroCORDEX-UK (Use and understanding of EuroCORDEX data over the UK)

This project aims to extend the current suite of UKCP climate projections by incorporating information predominantly from the Euro-CORDEX downscaling experiment.

Richard Chandler, Claire Barnes and Chris Brierley (University College London)

https://www.ukclimateresilience.org/projects/use-and-understanding-of-eurocordex-data-over-the-uk/

https://github-pages.ucl.ac.uk/UKCORDEX-plot-explorer/

Met Office—Improving Climate Hazard Information, from Climate Hazard to Climate Risk

ExSamples (Extreme samples)

This project aims to better understand the sampling statistics of extreme events in three winters predicted to be the hottest or wettest in UKCP18 global projections.

David Wallom (University of Oxford), with Met Office, University of Bristol

https://www.ukclimateresilience.org/projects/exsamples-extreme-samples/

Met Office—Improving Climate Hazard Information, from Climate Hazard to Climate Risk

FREEDOM-BCCR (Forecasting risk of environmental exacerbation of dissolved organic matter—building climate change resilience)

This project aims to predict climate impacts on the water industry, to inform best practice climate resilience strategies.

Don Monteith (UK Centre for Ecology & Hydrology), with University of Leeds, University of Glasgow, Scottish Water, United Utilities, Welsh Water, Yorkshire Water

https://www.ukclimateresilience.org/projects/freedom-bccr-forecasting-risk-of-environmental-exacerbation-of-dissolved-organic-matter-building-climate-change-resilience/

Phase 1 Projects

Freshwater Monitoring and Forecasting (Delivering resilience to climate impacts on UK freshwater quality: Towards national-scale cyanobacterial bloom monitoring and forecasting)

This project aims to demonstrate the power of new satellite data for monitoring algal blooms in waterbodies across the UK.

Peter Hunter (University of Sterling), with UK Centre for Ecology and Hydrology, University of Glasgow, Plymouth Marine Laboratory

https://www.ukclimateresilience.org/projects/delivering-resilience-to-climate-impacts-on-uk-freshwater-quality-towards-national-scale-cyanobacterial-bloom-monitoring-and-forecasting/

Phase 1 Projects

FUTURE-DRAINAGE (Ensemble climate change rainfall estimates for sustainable drainage)

This project aims to update guidance for drainage design and urban surface water flood risk assessment in the UK.

Hayley Fowler (University of Newcastle), with Loughborough University, Southern Water, United Utilities, Thames Water, JBA Consulting, Welsh Water, Scottish Environment Protection Agency, Anglian Water, Yorkshire Water

https://www.ukclimateresilience.org/projects/future-drainage-ensemble-climate-change-rainfall-estimates-for-sustainable-drainage/

Phase 1 Projects

Health Sector Resilience (Prototype development: Addressing the resilience needs of the UK health sector)

This project aims to characterise extreme events linked to ill-health in the UK and quantify how climatic and demographic changes might necessitate resilience in health and social care in future decades.

Andrew Charlton-Perez (University of Reading)

https://www.ukclimateresilience.org/projects/addressing-the-resilience-needs-of-the-uk-health-sector-climate-service-pilots/

Met Office—Pilot Climate Services

Improving Climate Hazard Information (Improving climate hazard information)

This project aims to produce a method to estimate changes in the probability density function of extreme value statistics compatible with the UKCP18 approaches.

Simon Brown, Peter Stott, Lizzie Kendon, Rob Shooter, William Keat, Daniel Cotterill, James Pope (Met Office)

https://www.ukclimateresilience.org/projects/improving-climate-hazard-information/

Met Office—Improving Climate Hazard Information, from Climate Hazard to Climate Risk

London Climate Action (Climate action strategy for City of London—Adaptive design/pathways for London's cubic mile)

This placement aims to support the delivery of the City of London Corporation's Climate Action Strategy 2020–2027.

Katy Freeborough (British Geological Survey), with City of London Corporation

https://www.ukclimateresilience.org/projects/climate-action-strategy-for-city-of-london-adaptive-design-pathways-for-londons-cubic-mile/

Embedded Researcher Cohort 2

MAGIC (Mobilising adaptation—governance of infrastructure through co-production)

This project aims to demonstrate and evaluate a community-led approach to reducing flood risk.

Liz Sharp (University of Sheffield), with University of Hull, Queen Mary University of London, Living with Water Partnership, Hull and East Riding Timebank

https://www.ukclimateresilience.org/projects/magic-mobilising-adaptation-governance-of-infrastructure-through-co-production/

Governing Adaptation

Manchester Climate Action (Adaptation and Resilience: Planning and action for Manchester)

This project aims to establish a policy and action-planning framework to enable Manchester sectors to take urgent and sustained action to increase resilience to climate variability.

Paul O'Hare (Manchester Metropolitan University), with Manchester Climate Change Agency
https://www.ukclimateresilience.org/projects/adaptation-resilience-planning-action-for-manchester/
https://www.manchesterclimateready.com/
Embedded Researcher Cohort 1

Meeting Urban User Needs (Prototype development: Meeting urban user needs)
This project aims to understand user needs for specific applications of local decision-making in aspects such as health, infrastructure and water.
Claire Scannell and Victoria Ramsey (Met Office)
https://www.ukclimateresilience.org/projects/prototype-development-meeting-urban-user-needs/
https://www.metoffice.gov.uk/research/approach/collaboration/spf/ukcrp-outputs
Met Office—Pilot Climate Services

Multiple Hazards (Multiple hazards and compound events)
This work aims to characterise risks from multiple climate hazards and how they may change in terms of location, severity, frequency and duration throughout the twenty-first century.
Dan Bernie and Freya Garry (Met Office)
https://www.ukclimateresilience.org/projects/from-climate-hazard-to-climate-risk/
Met Office—Improving Climate Hazard Information, from Climate Hazard to Climate Risk

National Framework for Climate Services (Facilitating the delivery and use of climate services)
This project aims to engage with climate services stakeholders to determine whether there is a need for a UK National Framework of Climate Services.
Natalie Garrett, Louise Wilson and Nicola Golding (Met Office)
https://www.ukclimateresilience.org/projects/facilitating-the-delivery-and-use-of-climate-services/
Met Office—Operational Climate Services

Once Upon a Time (Once upon a time in a heatwave)
This project aims to explore storytelling to communicate impacts of—and adaptation to—a changing climate in Northern Ireland, with a particular focus on agricultural and rural communities.

Alan Kennedy-Asser (University of Bristol), with Climate Northern Ireland

https://www.ukclimateresilience.org/projects/once-upon-a-time-in-a-heatwave/

Embedded Researcher Cohort 2

OpenCLIM (OpenCLIM: Open climate impacts modelling framework)

This project aims to support UK assessment of climate risk and adaptation by developing and applying an integrated assessment model.

Robert Nicholls (Tyndall Centre, University of East Anglia), with Newcastle University, University of Bristol, Science and Technology Facilities Council, UK Centre for Ecology and Hydrology, Environment Agency, Climate Ready Clyde, Broads Authority, Anglian Water, Arup, RMS, MottMacDonald, Universiteit Utrecht, Paul Sayers and Partners

https://www.ukclimateresilience.org/projects/openclim-open-climate-impacts-modelling-framework/

Enhancing Climate Change Risk Assessment

Resilience for Churches (Co-developing resilience strategies for churches and their communities)

This project aims to co-develop climate resilience in church buildings across the UK.

Chris Walsh (University of Manchester), with Church of England

https://www.ukclimateresilience.org/projects/co-developing-resilience-strategies-for-churches-and-their-communities/

Embedded Researcher Cohort 2

RESIL-RISK (RESIL-RISK: Understanding UK perceptions of climate risk and resilience)

This project aims to investigate how people currently conceptualise the relationship between climate risk, resilience and adaptation policy, as evidence for designing future communications.

Nick Pidgeon (Cardiff University), with Climate Outreach

https://www.ukclimateresilience.org/projects/resilrisk-understanding-uk-perceptions-of-climate-risk-and-resilience/

Phase 1 Projects

Risk Assessment Frameworks (Comparison of risk assessment frameworks)

This project aims to establish how to best assess climate change risk in support of climate resilience efforts, by identifying and comparing UK specific risk assessment frameworks.

Dan Bernie, Laura Dawkins and Kate Brown (Met Office)
https://www.ukclimateresilience.org/projects/from-climate-hazard-to-climate-risk/
Met Office—Improving Climate Hazard Information, from Climate Hazard to Climate Risk

Risky Cities (Living with water in an uncertain future climate)
This project aims to develop research-informed learning histories for flood-prone Hull for use in community-based arts and heritage interventions and large-scale productions.
Briony McDonagh (University of Hull), with Absolutely Cultured, Hull City Council, Hull Minster, Hull: Yorkshire's Maritime City, Living with Water Partnership, National Youth Theatre
https://www.ukclimateresilience.org/projects/risky-cities-living-with-water-in-an-uncertain-future-climate/
https://riskycities.hull.ac.uk/
Living with Uncertainty

SEARCH (SEARCH: Sensitivity of estuaries to climate hazards)
This project aims to evaluate climate flooding hazards in UK estuaries.
Peter Robins (Bangor University), Hull University, British Geological Survey
https://www.ukclimateresilience.org/projects/search-sensitivity-of-estuaries-to-climate-hazards/
https://www.researchgate.net/publication/363166162_Historic_Spatial_Patterns_of_Storm-Driven_Compound_Events_in_UK_Estuaries/link/631009675eed5e4bd136f581/download
Present and Future Climate Hazard

Stochastic Simulation (Facilitating stochastic simulation for UK climate resilience)
This project aims to explore how weather generators could be more widely used to support climate resilience activities in the UK, especially in flood and water management projects.
David Pritchard (Newcastle University), with JBA Consulting
https://www.ukclimateresilience.org/projects/facilitating-stochastic-simulation-for-uk-climate-resilience/
Embedded Researcher Cohort 2

STORMY-WEATHER (STORMY-WEATHER: Plausible storm hazards in a future climate)

This project aims to use the latest climate projections to develop a new methodology to understand what drives changes in extreme rainfall and windstorms for different storm types.

Hayley Fowler (Newcastle University), with University of Exeter

https://www.ukclimateresilience.org/projects/stormy-weather-plausible-storm-hazards-in-a-future-climate/

Present and future climate hazard

Time and Tide (Time and tide: Resilience, adaptation and art)

This project aims to investigate how the arts can catalyse communities to act—and become more resilient—as climate change intensifies and socioeconomic inequalities increase.

Corinna Wagner (University of Exeter), with Time and Tide Bell

https://www.ukclimateresilience.org/projects/time-and-tide-resilience-adaptation-art/

Embedded Researcher Cohort 2

Tourism Adaptation (Climate change and the tourism sector: impacts and adaptations at visitor attractions)

This project aims to investigate the relationship between weather/climate and visitation to heritage attractions under current and future climatic conditions.

Tim Coles (University of Exeter), with National Trust, Historic Environment Scotland

https://www.ukclimateresilience.org/projects/climate-change-and-the-tourism-sector-impacts-and-adaptations-at-visitor-attractions/

Embedded Researcher Cohort 2

Transport/Energy Climate Services (Climate services for the transport and energy sectors)

This project aims to scope and co-develop initial prototype climate services for the UK's transport and energy sectors.

Erika Palin and Kate Brown (Met Office), with Department for Transport

https://www.ukclimateresilience.org/projects/climate-services-for-the-transport-and-energy-sectors/

Met Office—Pilot Climate Services

UKCR Synthesis (Linking to the national Climate Change Risk Assessment (CCRA) process and synthesis of SPF resilience projects)

This project aims to ensure consistency with the CCRA3 and proper co-development of science.

Peter Stott, Richard Betts, Simon Brown and Elizabeth Kendon (Met Office)
https://www.ukclimateresilience.org/projects/improving-climate-hazard-information/
Met Office—Improving Climate Hazard Information, from Climate Hazard to Climate Risk
UK-SSPs (UK socioeconomic scenarios for climate research and policy)
This project aims to produce UK-specific downscaled socioeconomic narratives and gridded data for a range of indicators, extended to 2100.
Jon Stenning (Cambridge Econometrics), with UK Centre for Ecology and Hydrology, University of Edinburgh, University of Exeter
https://www.ukclimateresilience.org/projects/uk-socioeconomic-scenarios-for-climate-research-and-policy/
Met Office—Improving Climate Hazard Information, from Climate Hazard to Climate Risk
Upscaling Climate Service Pilots (Upscaling of climate service pilots into routine services)
This project aims to consider how existing UKCR-developed pilot services can be up-scaled to routine services, making them useful and accessible to stakeholders.
Galina Guentchev, Erika Palin and Christopher Goddard (Met Office)
https://www.ukclimateresilience.org/projects/upscaling-of-climate-service-pilots-into-routine-services/
Met Office—Pilot Climate Services
Water Sector Resilience (Towards forecast-based climate resilience and adaptation in the water sector)
This project aims to understand how improved forecast capabilities can inform future operations adaptation in the water sector in response to climate change and population growth.
Charles Rougé (University of Sheffield), with Anglian Water
https://www.ukclimateresilience.org/projects/towards-forecast-based-climate-resilience-and-adaptation-in-the-water-sector/
Embedded Researcher Cohort 1
Yorkshire Climate Action (Whose role is it to act on climate resilience? Implementing Yorkshire's Climate Action Plan with Leeds City Council)

This project aims to assist Leeds City Council's flood risk management team in developing a stronger leadership role on climate resilience and adaptation planning.

Stephen Scott-Bottoms (University of Manchester), with Leeds City Council, Flood Risk Management

https://www.ukclimateresilience.org/projects/whose-role-is-it-to-act-on-climate-resilience-implementing-yorkshires-climate-action-plan-with-leeds-city-council/

Embedded Researcher Cohort 2

Index

A
ACCELERATED, 139
Accessibility, 28, 136, 173, 192
Action research, 32
Adaptation, 2, 4–9, 29, 30, 47–49, 57, 64–66, 68–70, 72, 73, 79, 82, 96, 106, 111–115, 119, 121, 122, 135–137, 146, 147, 167, 169, 173, 174, 178, 180–182, 188, 189, 191, 195, 196
Adaptation Reporting Power (ARP), 5, 6, 106
Agriculture, 48, 115, 137, 140, 147, 152, 153, 154, 165, 166, 170, 172, 173
Anglian Water, 48
AquaCAT, 167–170
AquaCAT flooding event sets, 170
ARID, 47
Arts, 4, 8, 30, 32, 68, 76, 77, 79–81, 83, 84, 86, 188, 189
Arts and Humanities Research Council (AHRC), 8
Applied research, 96

B
Barriers, 4, 27, 28, 33, 68, 69, 72, 83, 103, 107, 124, 136, 195
Belfast, 113, 123, 170
Bristol, 31, 47, 48, 77, 120, 138, 178
Bristol City Council, 31, 47, 117, 178, 180
Bristol Heat Resilience, 47, 98, 117
British Geological Survey, 48

C
CAESAR-Lisflood, 136
Cascading risk, 12, 164, 196
Case studies of agricultural compound hazards, 134
Catastrophe modelling framework, 191
Church of England, 49, 193
City, 6, 31, 33, 70, 71, 73, 78, 84, 113, 117, 119, 123, 125, 166, 170, 178, 183, 190
City of London Climate Action Strategy, 48, 69
City of London Corporation, 48

216 INDEX

City Packs, 117, 119, 121, 122, 125
City scale, 68, 70
CLandage, 31, 67, 77, 78, 80, 81
ClimaCare, 66, 167
CLIMADA, 140, 168, 171
Climate Change Act 2008, 5, 106, 182
Climate Change Committee (CCC), 18, 115, 187, 191
Climate Change Risk Assessment (CCRA3), 115, 147, 156, 180, 189, 191
Climate hazards, 4, 70, 113, 119, 132, 136, 137, 139, 140, 147, 149, 156, 157, 158, 164–167, 190
Climate information, 29, 33, 45, 57, 70, 94, 100, 133, 140, 171, 172, 181
Climate Information for Decision Making, 70
Climate Northern Ireland, 6, 48, 114, 118
Climate opportunities, 7
Climate Outreach, 183
Climate Ready Clyde, 115
Climate resilience, 8, 9, 11, 20, 21, 28, 30, 36, 37, 48, 49, 57, 65, 66, 71, 76, 77, 83, 84, 86, 98, 99, 113, 124, 164, 174, 183, 186, 187, 189, 195–197
Climate Resilience Standards, 98
Climate resources, 147
Climate risk, 2, 7, 8, 11, 47, 48, 54, 57, 71, 79, 96, 114, 115, 117, 119, 121, 140, 147, 148, 156–158, 165–167, 169, 170, 173, 180, 188–191, 195, 196
Climate Risk Indicators, 114, 119, 121, 122, 149–153, 156, 166–168, 170
Climate scenarios, 114, 148, 158, 169

Climate services, 4, 8, 9, 11, 29, 31, 33, 35, 94, 96, 98, 99, 101–108, 139, 181, 182, 189, 190, 195, 196
Climate Services Delivery, 93
Climate Services Standards, 99
Climate Stress Testing, 47, 98
Coastal Climate Services, 96, 116, 119, 121, 122, 170, 171
CoastalRes, 115
Co-development, 99, 102, 104, 106, 180, 186, 196
Collaboration, 48, 50, 52, 53, 67–69, 71, 79, 104, 122, 141, 189, 194
Communicating risk, 168, 190
Community, 9, 11, 16, 17, 19, 20, 28, 30–33, 49, 64, 67–69, 71, 72, 76–79, 81, 83, 84, 86, 99, 104, 113, 169, 173, 178, 186–189, 195–197
Community engagement, 29, 81
Community-led, 36
Compound risk, 12
Co-production, 11, 28–38, 93, 121, 122, 124, 166, 186, 191–193, 195
Cornwall, 79, 81
COTIDAL, 81
Coupled Model Intercomparison Project (CMIP6), 139
Coupled Model Intercomparison Project Phase 5 (CMIP5), 116
COVID-19, 31, 37, 38, 44, 46, 104, 187
Creative Climate Resilience, 30–32, 68, 69, 77, 78, 81, 83, 84, 98, 183
Credibility, 28
CREWS-UK, 149, 182
Cumbria, 67, 78
CS-N0W, 139

D

Data explorer webpage, 134
DECIPHeR, 133
Decision making, 11, 12, 29, 30, 32, 35, 45, 52, 56, 63, 67, 72, 82, 106, 119, 121, 122, 132, 135, 140, 147, 173, 191
Decision support, 7, 106
Decision support tools (DSTs), 6, 11, 112–117, 121–125, 190
Degree days, 149, 152, 153, 156
Demand-driven, 122
Department for Business, Energy & Industrial Strategy (BEIS), 139
Department for Digital, Culture, Media & Sport, 84
Department for Education, 47, 114
Department for Environment, Food & Rural Affairs (Defra), 7, 96, 115, 140, 191
Department for Infrastructure Rivers (Northern Ireland), 116
Department for Transport, 97
Design rainfall flood uplifts, 134
Dialogue, 48, 51, 54, 56, 76–81, 83, 86, 178–181
Drought, 31, 47, 70, 97, 119, 139, 140, 146, 147, 149, 151, 153, 154, 155, 165, 170, 190

E

East Riding of Yorkshire, 78
Effectiveness or evaluation, 2, 9, 11, 44, 54, 56, 76, 83, 86, 95, 108, 122, 123, 141, 181, 188, 192
Embedding, 11, 44, 46, 52, 56, 57, 167
Enabling environment, 21, 95, 101, 103, 105
Engagement, 8, 11, 19, 32, 33, 35, 36, 51, 68, 78, 82–84, 86, 93, 96, 102, 104–106, 121–125, 166, 181, 192, 193
England, 6, 115, 154–156
Environment Agency, 47, 96, 97, 115, 116, 138
Environment Agency Incident Response, 47, 166
Essex, 79, 81
EuroCORDEX-UK, 134, 140, 171
European Roadmap on Climate Services, 94
Exposure, 12, 17, 19, 20, 106, 112, 132, 140, 141, 158, 163–167, 169, 170, 173, 191
ExSamples, 133–135
Extreme events, 136, 137, 153, 154, 168
Extreme heat, 155
Extreme precipitation, 133, 137, 138, 140
Extreme weather, 67, 72, 81, 165, 167, 180
Extreme winds, 139, 156
Extreme winter scenarios, 133, 134

F

Flood, 31, 32, 36, 47, 48, 67, 68, 70, 78, 80, 81, 84, 96, 114–116, 119, 133, 135, 136, 183, 189, 190
FloodLights, 81, 84
FREEDOM-BCCR, 97, 151
Freshwater Monitoring & Forecasting, 97
Friends of Par Beach, 79
Front identification code, 134
FUTURE-DRAINAGE, 96, 134, 138, 150, 156
Future of the Northern Irish Countryside, 79

G
Geographic information system (GIS), 30, 116, 117, 121
Glasgow, 83

H
Harwich, 79, 81
Hazards, 12, 70, 71, 113, 115, 117, 119, 122, 131–133, 135–138, 140, 141, 147, 148, 153, 156, 157, 166, 167, 170, 172, 190, 191, 196
Health, communities and the built environment, 147, 151, 155
Health Sector Resilience, 96, 152
Heat Resilience Plan, 47, 117
Heat Service, 113, 117, 119–121, 123
Heat Vulnerability Index, 47, 117, 170
High resolution modelling, 4
Historic England, 31
Hull, 36, 78, 80, 84, 113, 183, 188
Hull and East Riding Timebank, 36
Hull City Council, 84
Humanities, 4, 8, 76, 77, 80, 83, 84, 86, 188, 189
Hybrid working, 46

I
Impact, 4–6, 35, 46, 55, 65, 70–72, 77, 78, 82, 83, 86, 95, 98, 101, 103, 104, 106, 112, 113, 115, 117, 119, 121–123, 132, 133, 135, 137, 138, 140, 141, 146, 147, 153–158, 165–172, 178, 180, 181, 188, 190, 191, 193, 194, 196
Inequality, 20, 64, 72
Infrastructure, 4, 18, 67, 69, 70, 72, 96, 113, 114, 117, 122, 123, 135, 146, 147, 150, 154, 155, 165
Institution of Mechanical Engineers, 116
Interdisciplinarity, 52
Intergovernmental Panel on Climate Change (IPCC), 17, 132
Intergovernmental Panel on Climate Change's 5th Assessment Report (IPCC AR5), 116
Isle of Lewis, 79

J
JBA Consulting, 48

K
Knowledge-deficit model, 29

L
Lancashire, 81
Leeds City Council, 48
Levelling Up Bill, 84
Local, 6, 11, 20, 21, 29–32, 65–73, 78–82, 84, 86, 96, 98, 112–117, 119, 121, 122, 133, 136–138, 140, 141, 148, 149, 156, 157, 169, 171, 173, 178, 183, 191, 195
Local Flood Risk Management Strategy for 2022–2028, 84
London, 66, 188
London Climate Action, 48, 69
London Climate Change Partnership, 6, 47

M
MAGIC, 34, 36, 67, 68
Manchester, 47, 68, 70, 78, 180, 183
Manchester Climate Action, 47, 70, 166

Manchester Climate Change Agency, 47, 71
Meeting Urban User Needs, 70, 97, 113, 117, 119–121, 123, 166–168, 170, 172, 178
Met Office, 4, 5, 7–9, 11, 139, 155, 186, 187, 194
Miles Platting, 78, 79
Ministry of Justice, 114
Modelling, 7, 79, 100, 115, 119, 125, 135, 136, 167, 191
Morecambe, 81
Multidisciplinarity, 7, 36
Multiple Hazards, 134, 137, 139, 149, 150, 153, 166–168, 170, 172, 182

N
National Adaptation Programme (NAP), 5, 6, 180, 194
National Trust, 49
National Youth Theatre, 34, 82
Natural England, 115
Natural environment and assets, 148, 149, 153
NatureScot, 97
Newton Heath, 78, 79
NHS, 96
Norfolk Broads National Park Authority, 115
North Yorkshire, 81
Northern Ireland, 6, 71, 116–118
Northern Ireland Rural Heat Map, 114

O
Ofwat, 97
Once Upon a Time, 48, 71, 77, 79, 81, 114, 117, 118, 122
On the Edge, 77, 82

OpenCLIM, 96, 115, 119, 121, 125, 152, 167–170, 196
Outcomes, 31, 34, 44–46, 55, 76, 78, 83, 86, 95, 103, 122, 148
Outer Hebrides, 78, 81
Overheating, 66, 155, 173

P
Place, 5, 6, 11, 18–21, 30, 65–73, 76–81, 84, 86, 107, 147, 153, 196
Place-based, 18, 64
Pooley Bridge, 67
POSTNOTE 647 on Coastal Management, 84
Product development, 96, 97
Prototype development, 96, 97
Public health, 96, 138
Public policy, 70

Q
Qualitative mapping, 166

R
Rainfall, 4, 96, 113, 114, 119, 133, 135, 136, 138, 141, 149, 150, 153, 154, 156, 157, 172, 190
Redcar, 81
RED-UP Rainfall Perturbation Tool, 134
Regional, 5, 6, 67, 78, 79, 94, 114, 116, 121, 122, 125, 133, 137, 138, 140, 141, 148, 149, 156, 157, 169, 171, 172, 190, 195
Regulation, 20, 96, 102, 103, 106
Relevance, 11, 28, 53, 55, 183, 187
Representative Concentration Pathways (RCPs), 116
Research, 4–9, 11, 12, 28–38, 44–46, 50, 53, 55, 56, 65, 71, 75–83,

86, 94, 95, 99, 101, 106, 112, 122, 124, 137, 139, 148, 164, 166, 172, 173, 178, 180–182, 186–197
Resilience for Churches, 49
RESIL-RISK, 180
Return levels at high resolution, 134
Risk Assessment Frameworks, 114, 121, 140, 167, 168, 170
Risk indicators, 168
Risk of compound flooding map, 134
Risks, 5, 8, 11, 18, 37, 47, 66, 68, 96, 106, 112, 114, 146, 155, 157, 164–167, 169–171, 180, 181, 183, 188, 190, 195, 196
Risky Cities, 34, 77, 78, 80, 83, 84, 183
River catchment resilience, 133
Royal Geographical Society, 77

S
Scalability, 95, 104, 105
Science-driven, 122
Scottish Environment Protection Agency (SEPA), 96, 116, 138
Scottish Water, 97
SEARCH, 134, 135
Space4Climate, 47
Staffordshire, 78, 81
Staffordshire Record Office, 31, 32
Statistical methods for extremes, 169
Stochastic Simulation, 48, 134, 136
Stochastic weather generator, 48, 134, 136
Storm type dataset, 134
STORMY-WEATHER, 133, 134, 137, 139
Storytelling/Folklore, 31, 69, 71, 72, 79, 81, 83
Susdrainable, 68
Systemic risk, 12, 164, 196

T
Tailoring, 65, 187
Tasglann nan Eilean Siar, 31
Task Force on Climate-related Financial Disclosures (TCFD), 103, 106, 182
Threshold-based methodology, 166
Time and Tide, 49, 77, 79, 81, 183
Time and Tide Bell, 49
Tourism Adaptation, 49
Transdisciplinarity, 194
Transport/Energy Climate Services, 32
26th UN Climate Change Conference of the Parties (COP26), 82

U
UK Climate Projections 2018, 4, 178
UK Climate Projections 2018 (UKCP18), 4, 12, 113, 114, 119, 133, 136, 137, 139, 140, 147–149, 157, 169–172, 178, 190
UKCP Local Projections, 133
UKCP Regional projections, 133
UK Climate Resilience Programme (UKCR), 7–9, 11, 12, 15, 16, 19, 27–29, 33, 34, 36–38, 43, 45, 56, 66, 75, 77, 78, 82, 84, 86, 94, 96, 102, 106, 107, 112–114, 116, 118, 121–125, 131, 133, 134, 136–140, 145, 147, 148, 157, 166–171, 178–182, 186–192, 194, 196
UK Health Security Agency, 96
UK Heat Stress Vulnerability website, 156
UK Research and Innovation (UKRI), 7–9, 45, 187, 194
UK Shared Socioeconomic Pathways (UK-SSPs), 32, 99, 169–171, 173

Uncertainty, 125, 138, 140, 141, 146, 148, 153–157, 169–171, 190
UNESCO, 67
Upscaling Climate Service Pilots, 98, 125
Urban governance, 70
Urban Heat Service, 172
Urban planning, 138
Urban subsurface, 48, 69
Usability, 28, 95, 102, 105, 122, 124, 136, 181, 191
User trust, 93, 95, 104, 105

V
Vulnerability, 2, 12, 17, 19, 20, 32, 64, 72, 113, 117, 132, 140, 141, 147, 156–158, 164–167, 169, 170, 173, 191, 195

W
Water management, 48
Water Sector Resilience, 48, 97
Weather generator, 136
Weather systems, 133, 137, 141
Welsh Government, 156
Welsh Water, 123
Wildfire, 146, 150, 152, 153, 157, 166
The Wildlife Trust, 114, 156
World Bank, 99
World Meteorological Organization, 99, 196

Y
Yorkshire and Humber Climate Commission, 48
Yorkshire Climate Action, 48

Milton Keynes UK
Ingram Content Group UK Ltd.
UKHW010831010224
437054UK00006B/177